T0325122

The Business of Mining

The Business of Mining

Series Editors:

Odwyn Ifan Jones
Emeritus Professor, Curtin University, Perth, Australia

A.J.S. Spearing & Eric Lilford
Western Australian School of Mines, Curtin University, Perth, Australia

ISSN print: 2640-463X
ISSN online: 2640-4648

Volume 1

The Business of Mining

The Mining Business, Uncertainty, Project Variables and Risk, Royalty Agreements, Pricing and Contract Systems, and Accounting for the Extractive Industry

VOLUME 1

Odwyn Jones
Emeritus Professor, Curtin University, Perth, Australia

Eric Lilford
Western Australian School of Mines, Curtin University, Perth, Australia

A.J.S. Spearing
Western Australian School of Mines, Curtin University, Kalgoorlie, Australia

Grantley Taylor
Curtin Business School, Curtin University, Perth, Australia

CRC Press is an imprint of the
Taylor & Francis Group, an informa business
A BALKEMA BOOK

CRC Press/Balkema is an imprint of the Taylor & Francis Group, an informa business

Typeset by Apex CoVantage, LLC

Exclusive Licence to publish granted by Curtin University to CRC Press/Balkema

Library of Congress Cataloging-in-Publication Data
Names: Jones, Odwyn, author.
Title: The mining business, uncertainty, project variables and risk, royalty agreements, pricing and contract systems, and accounting for the extractive industry / Odwyn Jones, Former Principal of the Western Australian School of Mines, Kalgoorlie, Australia and Former Dean of Mining and Mineral Technology, Curtin University, Perth, Australia, Eric Lilford, Western Australian School of Mines, Curtin University, Perth, Australia, A.J.S. Spearing, Western Australian School of Mines, Curtin University, Kalgoorlie, Australia, Grantley Taylor, Curtin Business School, Curtin University, Perth, Australia.
Description: Leiden, The Netherlands : CRC Press/Balkema, [2019] | Series: The business of mining ; volume 1 | Includes bibliographical references.
Identifiers: LCCN 2018054280 (print) | LCCN 2018054838 (ebook) | ISBN 9781351173728 (ebook) | ISBN 9781138299689 (hardcover : alk. paper)
Subjects: LCSH: Geology, Economic. | Mineral industries. | Mines and mineral resources—Economic aspects. | Contracts.
Classification: LCC TN260 (ebook) | LCC TN260 .J66 2019 (print) | DDC 622.068—dc23
LC record available at https://lccn.loc.gov/2018054280

Published by: CRC Press/Balkema
Schipholweg 107c, 2316 XC Leiden, The Netherlands
e-mail: Pub.NL@taylorandfrancis.com
www.crcpress.com – www.taylorandfrancis.com

ISBN: 978-1-138-29968-9 (Hbk)
ISBN: 978-1-351-17372-8 (eBook)

Contents

Foreword by the vice-chancellor of Curtin University vi
Foreword by the editor viii
About the authors x

1 Overview of the mining business 1

2 Uncertainty, project variables and risk 12

3 Royalty pricing, agreements and contract systems in mining 29

4 Accounting for the extractive industry 59

Foreword by the vice-chancellor of Curtin University

The WA School of Mines, established in 1902, has been a core part of Curtin University since 1969 when it came on board to deliver mining education programs as part of what was then the Western Australian Institute of Technology.

As one of the first mining schools in Australia, it has adapted and expanded over more than a century to better support the education and research needs of the broader resources industry. In addition to retaining its mining roots, the School now also incorporates chemical engineering, mineral and energy economics and petroleum engineering.

Today, it holds a reputation as one of the best mining education and research centres nationally and internationally. The School was the top research performer in the 2017 QS international rankings, scoring 95 for citations per paper and 100 for H-index citations. In 2018, Curtin University was ranked second in the world for Mineral and Mining Engineering in the QS World University Rankings By Subject.

We are delighted to be bringing you this Focus Series, *The Business of Mining*, that captures the great wealth of knowledge that the WA School of Mines has to offer – in mine valuation and risk, orebodies and mineral exploration, accounting and mineral marketing.

The series' editor and original author Emeritus Professor Odwyn Jones – who himself led the WA School of Mines for 15 years – has done an outstanding job in harnessing this expertise and making it available to students and professionals in the global mining sector.

Curtin University's mining and minerals graduates are recognised and sought after around the world. Many go on to become high-achieving, dynamic leaders in their field.

Curtin's postgraduate program in mining and minerals is a rich and rewarding course of study: block teaching, online teaching and hybrid models of teaching have all been implemented to stay current, offer students flexible modes of study and ensure that field trips and on-site study are integral.

I am, therefore, especially delighted to see *The Business of Mining* align with Curtin's MOOC of the same name, and for royalties from the series to go towards supporting educational programs at the WA School of Mines and the mining leaders of tomorrow.

Professor Deborah Terry AO
Perth, Western Australia

Foreword by the editor

This series of Focus Books originated in a text written some twenty years ago. The intention was to publish it as a textbook with the royalties from sales being directed in their entirety to Curtin University's WA School of Mines.

Having been left to languish, the original text was finally submitted some three years ago to Professor Sam Spearing, following his recent appointment as Director of the WA School of Mines, for consideration and comment. He was of the view the material was worthy of being updated for publication and an approach was made to CRC Press/Balkema, Taylor and Francis Group, who agreed to publish the updated text.

Subsequently, a contract with the publisher was finalised, on the understanding that material would be published as a series of Focus Books, and that all royalties from sales be directed to the Curtin University to support undergraduate or postgraduate students wishing to study full time for one year or more at the WA School of Mines campus in Kalgoorlie.

The complete set of three Focus Books will provide readers with a holistic all-embracing appraisal of the analytical tools available for assessing the economic viability of prospective mines. The books were written primarily for undergraduate applied geologists, mining engineers and extractive metallurgists and those pursuing course-based postgraduate programs in mineral economics.

However, the complete series will also be an extremely useful reference text for practicing mining professionals as well as for consultant geologists, mining engineers or primary metallurgists.

Each volume has a discrete focus. Volume 1 presents an overview of the mining business, followed by an analysis of project variables and risk, an overall coverage of the royalty agreements, pricing and contract systems followed by a final chapter on accounting standards and practises for the minerals industry.

Volume 2 discusses, in some depth, alternative means of assessing the economic viability of mining projects based on the best estimate of the recoverable mineral and/or fossil fuel reserves.

Volume 3 commences with the 'JORC Code' – The Australasian Code for Reporting of Exploration Results, Mineral Resources and Ore Reserves. This volume also includes an introduction to the nature and origin of minerals, fossil fuels and orebodies, followed by a review of mineral exploration and sampling of mineral deposits. It concludes with a detailed section covering the basic principles and application of the various methods of estimating the in-situ mineral resources and ore reserves.

These Focus Books should also be particularly useful for students enrolled in Curtin University's MOOC 'The Business of Mining'. This course, designed in collaboration with leading industry and educational experts, concentrates on the theory and practice of appraising the economic merit and valuation of mining ventures.

My sincere thanks to my daughter Sian Flynne, Business Manager for Curtin's School of Economics and Finance, for her consistent support, professional advice and typing of the entire text. Sincere thanks are also due to all co-authors for their generous assistance in updating, and where necessary, extending the text to meet current day standards and demands. As co-editors, Dr Eric Lilford and Professor Sam Spearing have made invaluable contributions in reviewing all three volumes and providing overall editorial direction – in addition to being co-authors themselves.

The administrative procedures involved in dealing with all co-authors, university officers and the publisher were complete and made much easier by the friendly assistance and guidance of Ms Valerie Raubenheimer, Vice-President Corporate Affairs, and the assistance of Dr Andrea Lewis, publications consultant on behalf of the University.

Finally, a special word of thanks to the Curtin University's Vice-Chancellor, Professor Deborah Terry AO, for supporting the venture and to my colleague Professor Sam Spearing for his motivation, energy and friendship in making sure this project was initiated and completed on time. Lastly, but by no means least, sincere thanks to my long-suffering wife for her consistent support and wise counsel.

Emeritus Professor Odwyn Jones AO
Perth, Western Australia

About the authors

Odwyn Jones, Emeritus Professor commenced work as a mining trainee with the UK National Coal Board (NCB) in 1950 and was granted an Industry Sponsored Scholarship a few years later to read for a mining engineering degree at the University College of Wales, Cardiff, where he graduated with a BSc with first class honours. He then returned to the industry, obtaining his Colliery Manager's Certificate.

In 1957 Emeritus Professor Jones accepted the position of Lecturer in Mining Engineering at the Royal College of Science and Technology, Glasgow, which later became Strathclyde University. His part-time research, involving both laboratory work and field-testing at a local colliery, was sponsored by the NCB, and he graduated with a PhD from the University of Glasgow in the mid-sixties.

In 1970 he became Principal Lecturer in Environmental Technology of Buildings at the Polytechnic of the South Bank, London, before moving on to Bristol Polytechnic in 1973 as Head of Department of Construction and Environmental Health.

In 1976, Emeritus Professor Jones and his family moved to Western Australia where he took up the joint posts of Principal of the WA School of Mines, Kalgoorlie, and Dean of Mining and Mineral Technology at the WA Institute of Technology, which later became Curtin University.

In 1991 he transferred from Kalgoorlie to the University's main campus in Perth as Director University Development (International) and Director of the Brodie-Hall Research and Consultancy Centre.

Having retired from the University in 1995, he served as Visiting Professor at the Virginia Polytechnic Institute and State University, Blacksburg, for a term before assisting Perth's Central TAFE in developing its minerals and energy-related programs, where he stayed until 2001.

He is a longstanding Member of Engineers Australia, and Fellow of both the Australasian Institute of Mining and Metallurgy and the Institute of Materials, Minerals and Mining (UK).

He was Deputy Chairman (1981–98) and Chairman (1998–2013) of the Research Advisory Committee of the Minerals and Energy Research Institute of WA, a Statutory Authority operating under its own Act of Parliament.

His recent awards include an Officer of the Order of Australia in 2005 and the Australasian Institute of Mining and Metallurgy's Service Award for 2012–13. He was inducted into the Australian Prospectors and Miners Hall of Fame in September 2013.

Eric Lilford is a minerals economist and engineer, with interests covering quantitative research, advanced valuation methodologies, determining regulatory impacts on mineral and energy economics to both companies and governments, commodity trends, guiding mining and exploration companies, capital markets and developing and executing economic growth and optimisation strategies. Dr Lilford has multi-commodity and multi-jurisdictional expertise obtained over 29 years in the resources industry in technical, economic, financial, management and academic roles.

His current roles include lecturing and researching in minerals and energy economics at Curtin University and progressing early-stage studies for exploration companies focussed on sub-Saharan Africa. Dr Lilford retains the position of non-executive Chair of SSC, a South African-based commodities focused company. He has also held the chair of one and the managing director of three ASX-listed companies, as well as NED positions of numerous other listed companies. Prior to this, he was national head of mining and a corporate finance partner at Deloitte Touche Tohmatsu in Perth. Before emigrating to Australia, he was a director of project and business development at BSGR, where he managed aspects of large copper-cobalt projects, mines and plants located in Zambia and the DRC.

Dr Lilford has conducted extensive mining and investment business in more than 22 countries, has presented at numerous conferences world-wide and has authored many peer-reviewed topical articles in high quality journals. He holds a BSc and an MSc in Engineering, and a PhD in Mineral Economics. He is a Fellow of the Australasian Institute of Mining and Metallurgy.

Sam Spearing is currently Director of the WA School of Mines in Kalgoorlie, Curtin University and is active in rock mechanics, mine design, training and support research. He has co-invented several support products commonly used for support in tabular mining operations.

Professor Spearing received his BSc in Mining Engineering and Master of Civil Engineering from the University of the Witwatersrand. He completed a PhD in ultra-deep hard rock tabular mining, support and backfilling at the University of Silesia.

Professor Spearing held the position of Associate Professor in the Department of Mining and Mineral Resources Engineering at Southern Illinois

University Carbondale prior to joining Curtin in 2015 as Deputy Director of the WA School of Mines.

Professor Spearing is the author of the SAIMM-published book *Rock Mechanics*, has authored more than 40 peer-reviewed papers, supported more than $3.0 million dollars of funded research since joining academia in 2007, and been published in a variety of technical books and related publications.

Grantley Taylor has an honours degree in Science from the University of Adelaide and a PhD in Accounting from Curtin University. He commenced his career in the mining industry when he worked for North Limited/Geo-peko for 12 years in several operations throughout Australia.

Following a career in the mining sector, Professor Taylor then worked for the Australian Taxation Office for three years where he worked in the Large Business and International Sector, specifically the Energy and Resources Division, located in Perth. In 2006, he commenced teaching and research in the School of Accounting at Curtin University where he now works as professor and deputy head of that school. His research focuses on accounting as applied to the extractive industry, taxation, corporate governance, financial instruments, capital markets, corporate social responsibility. He has published widely in a range of international journals.

1 Overview of the mining business

'If it cannot be grown, it must be mined' (Unknown) is a truism that captures the importance of mining in our evolution and its ongoing need as civilisation moves ever forward. Our civilisation has been defined by mining:

- Stone Age – pre 4000 BC
- Bronze Age – 4000 BC to 1500 BC
- Iron Age – 1500 BC to about 1800 AD
- Steel Age – 1800 to about 1945
- Nuclear Age – post 1945.

Figure 1.1 shows our dependence on minerals in the United States. It is still valid qualitatively and is globally correct.

Mining is a key commercial business which involves the production, processing, waste disposal, purchase and sale of goods and the management of a wide range of operational and advisory services which are required to ensure a safe, sustainable and profitable outcome. Mining companies purchase or lease mineral deposits or orebodies and convert them into saleable commodities and waste products. The processes involved in the securing or purchasing of the raw material can vary significantly.

Deposits can be either greenfields (in areas where exploring for or mining the type of ore or mineral has not occurred previously) or brownfields (in areas where similar ore or minerals have already been exploited or explored) where the technical risk and financial reward are obviously reduced.

If the findings of an exploration investigation are positive and are acceptable to the board of directors, investors and lending institutions, the required finances will flow to allow the development of the mine and associated plant and infrastructure. Alternatively, a mining company may, through a joint venture or takeover of another company, purchase a share in or gain 'total control or ownership' of a proven orebody or an operating mine.

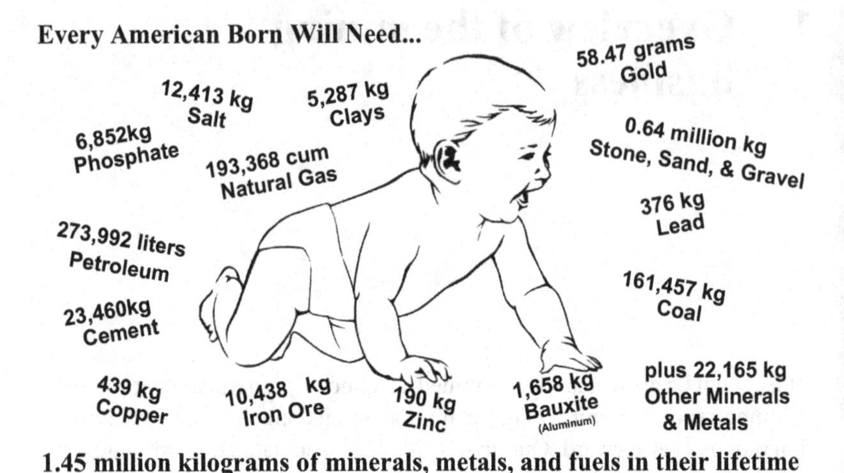

Figure 1.1 Human consumption over a lifetime

This chapter includes an overview of the factors which are taken into account in developing the feasibility and bankable documents required to bring a prospective mine into operation. It also deals with the valuations of mines and the assessment of mining projects.

Extractive industries

'Extractive industries' is a term used to define those industries involved in extracting minerals or fossil fuels from the earth's crust in order to produce material or products of use to mankind. It therefore embraces:

- Coal mining
- Oil and gas production
- Metalliferous mining
- Industrial minerals.

Extractive industries have always had an important influence on civilisation. They provide our major sources of energy and the raw material for the manufacture of almost all products which continue to shape our lives.

It is also important to remember that our mineral and fossil fuel deposits are wasting assets and if production is to be maintained, new replacement deposits have to be continually found. Intuitively, as the world's population continues

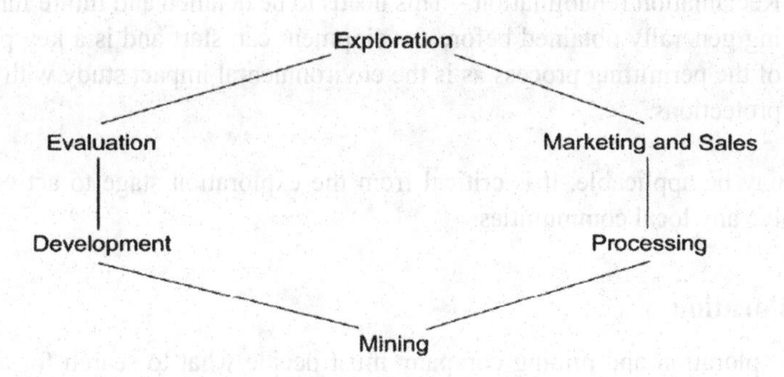

Figure 1.2 A typical exploration to mining cycle

to expand, merely replacing existing production rates will not sustainably satisfy the growing demands of the global economy. Therefore, the need exists to actually replace as well as grow exploration, or 'new find', rates.

The mining cycle therefore commences with exploration for viable deposits and is completed when the energy product, mineral or metallic commodity is marketed or sold and the waste safely disposed of. It is however important to point out that it is a continuous cycle with some of the income generated from commodity sales having to be reinvested in exploration in order to ensure the long-term viability and existence of the company.

Whilst this book concentrates on the coal and metalliferous mining industries, it includes references to the allied industries responsible for the production of oil and gas as well as industrial minerals.

The following are the stages involved in a mine life cycle:

- Literature review – This involves reviewing published data available from, say, the country's geological survey or equivalent and any data from previous exploration activity.
- Prospecting – This is the first stage of exploration and generally involves considerable hard work by geologists ('foot slogging' frequently).
- Exploration – Once a target location has been identified, the exploration of it to establish its viability is undertaken, keeping costs as low as possible in case it proves uneconomic. Prospecting and exploration are the domains of geologists and geophysicists.
- Development – Development involves the construction of the mine site including access, provision of services, any processing facility, the mine access and future waste handling/disposal systems.
- Exploitation – This involves the production of the saleable product from the mine. Development and exploitation are the domain of the mining engineer and metallurgist.

- Reclamation/rehabilitation – This needs to be detailed and future funding generally obtained before development can start and is a key part of the permitting process as is the environmental impact study with its protections.

As may be applicable, it is critical from the exploration stage to actively involve any local communities.

Exploration

An exploration and mining company must decide what to search for and where to search for it. In assessing what to explore for, the long lead time between the discovery of a deposit or an orebody and its exploitation must be taken into account and can vary from a few years to over 10 years. Consequently, those responsible must take a long-term view of the need for the mineral commodity and this entails taking into account:

- The likely future demand for the commodity, factoring in such unknowns as future disruptive technology trends and the possible use and/or development of alternative materials.
- The likely future supply of the commodity as existing producers may cost-effectively expand production or potential new producers may bring economically more attractive production onto the market.
- The local availability and cost of labour and services.
- The likely future trends in royalties and taxes, as well as requirements around local equity participation.
- The political and social stability.

Before opting for an exploration area, which could be in any part of the globe, it is important to assess:

- Sovereign risk
- The exploration potential of the area
- The likely costs of exploration and any subsequent mine development
- Access to the product market.

Assessing the geological potential of a prospective exploration area involves an examination of available geological maps and any available geochemical or geophysical surveys. Due recognition should also be given to the existence of operational mines and/or known mineralisation.

Exploration and mine development costs would be influenced by distances from major supply centres and the availability of power supply and transport services. Account should also be taken of the availability of skilled labour and management staff.

In assessing the political situation, account must be taken of all factors controlled by governments such as:

- Legislation governing exploration and mining activities.
- Policies governing corporate taxation and mineral royalties.
- Legislation governing equity participation by either one or both of local and government bodies.
- Regulations governing the funding of foreign owned business ventures.
- Ability to repatriate capital through dividends or other means unimpeded and without penalty.

Project risks

There are risks associated with any mining project. Before embarking on the more costly but vital diamond drilling, aerial photography, geological, geochemical and tomography studies are usually undertaken first. This preliminary exploration program will only reveal approximate estimates of the potential deposit or orebody and subsequent core drilling is essential to define the overall structure of the deposit, grade (and distribution) and surrounding strata. There are also diminishing levels of risk associated with the subsequent stages of mine planning and development due to the difficulty of designing cost effective and efficient extractive and treatment processes. These risks are reflected in the available sources of finance for the various stages in mine development.

Risk	*Activity*	*Source of Finance*
Very High	Exploration	Seed Capital
Very High	Scoping Study	Venture Capital
High	Pre-Feasibility Study	Venture Capital
High	Feasibility Study	Venture and/or Equity Capital
Moderate	Mine Development	Equity and/or Project Loan
Normal	Mine Production	Cashflows or Short-Term Loans

Project financing

Due to the very high risk associated with mineral exploration, its financing is either provided by operating mining companies seeking to extend their in situ reserves or by speculators who are prepared to invest their savings in grass-roots exploration with the expectation that success would reap a high reward. Once an attractive resource has been located, a scoping study or pre-feasibility study may be carried out to attract the venture capital required to carry out the costly and time-consuming exercise of completing the final feasibility study. The principal objectives of a scoping or pre-feasibility study are to:

- Gauge the market's economic attractiveness of the commodity being considered.
- Examine all possible project alternatives.
- Identify those aspects of the project that are critical to its feasibility.
- Explore whether the project proposal would be attractive to a particular investor or investor group in order to meet the expense of carrying out a detailed feasibility study.

For the various stages in determining the funding needed for a specific project, the following levels of accuracy are used as a general guideline (AusIMM, 2012):

- Conceptual design costs to ±30%
- Pre-feasibility costs to ±20%
- Feasibility costs to ±10%.

The scoping and both feasibility studies are structured in similar ways. The major difference is the much more detailed analysis of the technical, financial, economic and environmental factors contained in the final feasibility study report. Obviously this final study would either lead to a recommendation to continue or discontinue with the project. It should include an estimate of the working capital required and an investment appraisal and, if it recommends that the project proceed, it should identify the preferred sources of finance. The major sources of project finance are:

- Existing cashflows (if the company has other operations)
- Equity finance, including convertible loans
- Debt finance
- Commodity loans.

Equity finance is obtained from shareholders. As an example for a major mining and processing facility, the estimated total funds required were approximately $1.5 billion, of which $1.3 billion was the estimated capital cost. The proposed sources of funds included about $700 million of equity funds with most of the remainder being bank loans. Equity can be raised by issuing ordinary and/or preference shares. Ordinary shares have full voting rights, with dividends dependent on company profits. Preference shares, on the other hand, usually involve a dividend at least partly independent of profit, and with or without voting rights.

The return on investment expected by mine financiers varies significantly from about 8% to 18% depending on the uncertainty including factors such as local political stability; whether the mining company is a junior newcomer or has an established track record; and the market stability.

A convertible loan allows the project company to raise funds for development with the repayment of the loan replaced with a conversion of the principle loan amount into equity in the holding company, at a predetermined price.

Debt financing can be achieved through debentures (fixed interest securities), bank loans or overdraft. Whilst they are usually lower cost than equity funds, they carry a fixed charge and must be redeemed by a specified date. If things go well, debt financing can increase the return to the company. If, however, things go worse than expected, the high debt financing level can lead to bankruptcy due to the high minimum repayments.

A simplified example presented in a booklet entitled 'Will It Make a Quid?' by Viv Forbes demonstrates these issues. It deals with a mining project with a total capital cost of $100 million, which is expected to generate $10 million profit before tax and interest repayments.

- Case 1 defines the profile if the project is financed entirely by equity funds.
- Cases 2 and 3 demonstrate the effect of an increase and reduction of 20% in project profitability respectively.
- Case 4 defines the company profile if it has 90% debt finance.
- Cases 5 and 6 illustrate the effect, once again, of an increase and reduction of 20% in project profitability.

Whilst the returns to the owners (shareholders) in Cases 2 and 3 vary in proportion to changes in project profitability, a 20% increase in profitability in Case 5 results in a doubling of the return to shareholders. In Case 6, however, when the project profitability goes down by 20% the shareholders receive nothing and any further reduction in project profitability will lead to bankruptcy, since the owners would not be able to meet their interest payments.

	All Equity			*90% Debt*		
	Case 1 Base	*Case 2 +20%*	*Case 3 −20%*	*Case 4 Base*	*Case 5 +20%*	*Case 6 −20%*
Financing						
Equity	100	100	100	10	10	10
Debt	–	–	–	90	90	90
Total Capital	100	100	100	100	100	100
Profits						
Operating Profit Margin	10	12	8	10	12	8
Interest at 9%	–	–	–	8	8	8
Profit Before Tax	10	12	8	2	4	0
Tax at 50%	5	6	4	1	2	0
Profit After Tax	5	6	4	1	2	0
% on Equity	5%	6%	4%	10%	20%	0%

Commodity loans

In the 1960s and 1970s attracting investment for the development of new gold mines in Australia was extremely difficult and only those with very short payback periods were financed. That aside, prior to 1979 the Australian gold mining industry was in a perilous state, with total mine production less than 600,000 oz and dropping. From the 1980s, however, the gold price in Australian currency had more than trebled and the availability of carbon-in-pulp technology ensured that the near surface clay-rich ores could be treated at relatively low cost. At about the same time project financing based on 'gold loans' was developed in Australia.

This form of project financing was based on 'gold in the ground' being the major security for borrowing gold from a bullion bank to meet the cost of mine development. More recently, this method of financing has been extended to include other commodities such as copper.

Gold loan transactions are subject to gold loan agreements specifying the terms and conditions of the transaction, which include:

- The maximum and minimum amounts of gold that can be drawn down.
- The period of the gold loan, which is usually between two and five years.
- The arrangements for the re-delivery of gold to the lender.
- The gold borrowing fee, which was generally 1.5% to 2.5% in 2013 and is payable in gold.

The agreement may also dictate a gold trading strategy, whereby the borrower minimises the risk associated with a fluctuating commodity price by forward sales of a proportion of the gold production to cover the cost of production. If, for example, a mine produces 120,000 ounces of gold per annum at an operating cost of A$ 1,270 per ounce when the gold price is US$ 1,200 per ounce and the exchange rate is US$ 0.75 cents to an A$ (i.e. a gold price of A$ 1,600/oz), 80% of production would have to be sold forward at this price if the risk of falling spot prices is to be eliminated. In point of fact, the forward price is calculated at the spot price prevailing at the time plus the contango to cover interest and other charges. Intuitively if a contango (increasing forward price) does not exist and backwardation (decreasing forward price) does, locking in a forward price may not be optimal over a longer term, but may still be preferable over a shorter term, while the forward price remains above spot.

Feasibility studies

Whereas opportunity studies may be carried out during the exploration stage, the two major studies are the pre-feasibility and final feasibility (bankable or definitive) studies. The former is carried out to justify the expense involved in completing the much more detailed, costly and time-consuming work associated with the final feasibility study (a bankable document that can be used to try and obtain internal or external project funding). The main objectives of a pre-feasibility study, per Behrens and Hawranek (1991), are to:

- Examine all project alternatives and ascertain whether a detailed final feasibility study is justified.
- Identify those issues critical to the project's feasibility.
- Explore the project's environmental impact.
- Define the nature and extent of the exploration program, laboratory analysis and test-work required to provide the information needed for the final feasibility study.

Consequently, the pre-feasibility study would examine and report on:

- Available geological and geotechnical information.
- Estimation of mineral resources marketing and sales.
- Human resources.
- Environmental impact and management.
- Availability of and access to essential services (i.e. water, power and transport).
- Next stage in exploration, sample analysis and test-work and their associated costs.
- Mining and mineral processing methods (including safe waste disposal).
- Preliminary financial analysis and economic impact of relevant factors.

It is important to bear in mind that the analysis of mining (and oil and gas) projects differ from the analysis of most other engineering projects due to the fact that it is only possible to estimate the nature and content of the mineral resource, which is the project's major asset. It is paradoxical that the most reliable estimates are required in the pre-extraction stage when the least amount of information is available. This is, however, the time when feasibility studies have to be carried out and major investment decisions taken.

Considerable attention must, therefore, be devoted to estimating the most likely mineral reserves of all mining projects. Whilst this is a major issue, those involved in financing mining projects are exposed to the risk that any one or more of the following factors may not turn out as forecast:

- Mineral reserves
- Geological conditions
- Recovery factors
- Mining and processing costs and associated efficiencies
- Commodity sale prices
- Reclamation/rehabilitation costs.

The final feasibility study would be submitted to the board of directors for its consideration, the outcome being the continuance or otherwise of the project. A company frequently has more than one mining project being developed so will obviously select the one with the least risk and highest rate of return. A recommendation for continuance would be based, amongst other things, on a financial analysis of forecasted income and expenditure over the life of the project such that it has a positive net present value (NPV) and meets the targeted rate of return. It would identify the risks associated with the project and include a risk analysis assessing the sensitivity of the project's NPV to the project variables. Finally, it would indicate the preferred means of project finance and, once approved, would be submitted to the preferred bank, or lending institution, as justification for the granting of a loan (Rudeno, 2012).

Final feasibility studies should arrive at definitive conclusions on all aspects of the project. Such conclusions relating to mining systems, mineral processing, human resources, environmental management and project financing should be clearly explained and supported by sound arguments and compelling evidence. Whilst the format of these studies can vary, there is a core of material common to all, including:

- Corporate objectives
- Geology of the region with special reference to the mineral deposit and surrounding strata

- Location, site and environment
- Resource evaluation
- Legislative issues
- Mine development proposals and schedule
- Mining system and production schedules
- Mineral processing
- Availability and provision of water, power and site-specific transport services
- Infrastructure including road and rail logistics solutions to market
- Human resources (including their skills level)
- Markets, marketing and sales
- Socio-economic influences and impacts
- Financial analysis and investment appraisal.

As stated earlier, there are risks associated with any mining project, and a project loan can only be obtained if the final feasibility study proves beyond reasonable doubt that loan repayments can be met from the cash flows generated from mineral commodity sales. In analysing the veracity of the final feasibility study findings, all basic assumptions and estimates underlying the study should be questioned, including:

- Establishing the confidence which can be placed in the estimates of mineable reserves, grade, production levels, recovery factors, costs, sale prices and so forth.
- Confirming the existence or otherwise of sales contracts in draft form or as 'letters of intent' and the ability of the vendor to meet commodity specifications.
- Assessing the quality and extent of the financial analysis and investment appraisal, which should include identification of areas of uncertainty, critical variables and risks and address the means of risk management and explore probable future scenarios and their likely impact on the financial viability of the project.

References

Behrens, W. and Hawranek, P.M. 1991. *Manual for the Preparation of Industrial Feasibility Studies*. United Nations Industrial Development Organisation.

Forbes, V. 1980. *Will It Make a Quid*? A Discussion Paper presented to a Mining in Society Seminar, University Minerals Industry Advisory Council, Brisbane.

Guidelines for Technical Economic Evaluation of Mineral Industry Projects, AusIMM, 2012.

Rudeno, V. 2012. *The Mining 'Valuation' Handbook*. Melbourne: Wrightbooks Pty Ltd.

2 Uncertainty, project variables and risk

Uncertainty is unpredictability, or something that is not surely known, and those uncertain factors influencing a project's viability are known as project variables. Risk, on the other hand, is the chance of realising a sub-optimal or inefficient outcome or loss, such as the bankruptcy of a business or company. The level of uncertainty in a mining project is generally associated with the type and nature of the resource and the more the uncertainty surrounding this the higher the project risk.

Everyone lives in an uncertain world and face a multitude of risks in our daily lives. Boards of directors make business decisions in the light of the information available at that time produced by technical and financial experts. Their reports and recommendations will inevitably be based on a number of assumptions (best estimates) relating to the overall market, the specific project costs and the project's forecast earnings. The board's decision will be the collective subjective judgement of its members. It is, therefore, important for all decision makers to minimise the risks associated with their decisions and to mitigate those risks where possible.

The financial analysis of any mining project typically involves estimating its Net Present Value (NPV), which is the sum of the discounted or Present Values (PV) of all annual cash flows minus the project's cash outlay or capital cost, i.e.

$$\mathrm{NPV} = \sum_{n=:}^{n=n} \frac{CF_n}{(1+r)^n} - \mathrm{C}$$

Where: CF_n = Free cash flow at time n (typically years)
r = Rate of return (discount rate)
C = Capital cost of project or cash outlay.

Projects with a negative NPV are usually rejected.

Free cash flow = Cash receipts less cash disbursements.

If disbursements (costs) exceed receipts for the period in question, the cash flow for that period is negative.

The NPV of a project is a measure of the amount by which the project will increase the value of the business to its owners after taking into account that owner's required investment return on capital invested. In calculating the PV of an investment you need to know two things: the cash flow generated during each period (usually annually) over the life of the project and the minimum return expected from that investment. The discount rate is the return you would expect to make on other investments of comparable risk. If, for example, the discount rate is 10%, an investment of $90.91 today would generate $100.00 in one year's time. Taking the investment horizon a few years longer, Table 2.1 shows how the discount rate impacts forecast earnings.

Stated differently and as seen in the table above, if $56.45 was invested today over a six year period at an annual compounding return of 10%, that investment would be worth $100.00 at the end of the sixth year.

It is important to remember that PVs are market values and that the NPV of a project quantifies the extent to which an investment will add value to the company, because it measures the extent to which the investment's PV exceeds its cost. When boards of directors invest in a project with a positive NPV, they do so on the basis that they are making the shareholders wealthier. Companies can gain access to capital at the capital market rate and investments that promise a higher return than the capital market rate should make the shareholders better off. The capital market rate is typically a weighted rate incorporating a number of factors including the cost of borrowing attributed to equity and to debt. It also incorporates a country's specific, non-diversifiable risk (sovereign risk) as well as market volatilities.

The previous conclusion is, however, dependent on the accuracy of the Discounted Cash Flow (DCF) analysis, and care should be taken to examine

Table 2.1 Discounted cash flow values

Year		*1*	*2*	*3*	*4*	*5*	*6*
Future Income	$	100	100	100	100	100	100
Discount Factor	%	10%	10%	10%	10%	10%	10%
Present Value	$	90.91	82.64	75.13	68.30	62.09	56.45

its underlying assumptions and the accuracy of its estimates. The adage 'garbage in equals garbage out' is rather apt in cash flow modelling. Bearing in mind the extent of the unknowns associated with mining projects, such a critical appraisal is essential and special attention should be paid to the confidence levels which can be apportioned to the estimates of mineral production and associated yields, mining and mineral processing costs and the selling price of the mineral commodity. Depending on where the mining project is located, exchange rates may also have a significant impact on the final analysis since commodities are typically priced and traded in US dollars but the operating costs are expensed in the local currency.

It is, of course, vitally important that those responsible for preparing a final feasibility study of a proposed new mine minimise the uncertainty and risks associated with all estimates used in the financial analysis. Boards of directors should see evidence of this rigour in feasibility study reports.

The risks associated with the various stages in the development of a new mine influence the source of available funds as shown in Table 2.2.

Most banks are not interested in the speculative investment associated with exploration and evaluation because the risks are too high for their business model, and equity funds or strategic equity partners are the only sources of finance for these early stages. It is only when the development stage is reached that debt financing becomes a possibility. Since the cost of debt is generally lower than the cost of equity, it is intuitive then that the investment costs associated with exploration and evaluation, being the

Table 2.2 Funding sources dependent on development stage

Activity	*Risk*	*Funding*	*Source of Funds*
Exploration	Very high	Equity	Retained earnings, company float (IPO), rights issue, private seed or venture capital, farm-in or JVs
Evaluation	High	Equity, convertible debt	As for exploration sources
Development	Moderate	Limited recourse finance, corporate borrowing, and/or equity, derivative instruments	Trading banks, merchant banks, retained earnings, rights issue, share placement, JVs, hedging
Production	Normal	Non-recourse or limited recourse finance, corporate borrowing, and/or equity, derivative instruments	As for development stage, and including strategic investors (e.g. offtakers)

cost of securing capital for the early stages, will be higher than that for the development and production stages of mining.

At this later stage, project financing using limited recourse funding may be available, whereby the banks have full recourse to all the company's assets until full mine production is achieved. Thereafter the banks maintain a first ranking mortgage over the mine tenements and first ranking charge over other project assets. This means that the relevant bank has a first right to the assets in the event of a financial or economic default (default on repayment of borrowings or loss of profitability and hence asset closure or discontinuance). Other institutions may hold a second rank which means they are second-in-line for the assets if there is any value remaining after the first ranking creditors have balanced their overdue amounts. Financing later projects at an operating mine may attract non-recourse project finance, whereby the banks rely solely on the mine's cash flow to service the loan capital and interest with security confined to a mortgage and charge over all unencumbered project assets.

In addition to these securities, the banks may also require metal mining companies to have a price protection scheme in place to minimise or eliminate the downside of that specific commodity price's volatility. This can be achieved by forward selling a proportion of the mine's metal production to cover all or a substantial portion of the company's mining and processing costs. Generally, not more than 70% of any one year's forecast production is sold forward in this way.

It is, therefore, clear that banks take great care to reduce risk before granting project finance for a mining venture. The major risk is that revenue from the sale of the mineral commodity will be insufficient to meet the mine's operating costs and debt (capital plus interest) repayment. This could occur due to a blowout in costs caused by poor management or inappropriate mine and/or mineral processing plant design with poor throughput and recovery achievements. On the other hand, forecasted revenues may be unachievable due to lower production rates, lower ore grades, poor mineral recovery or efficiencies in the processing plant, or lower selling price of the mineral commodity.

The banks, therefore, adopt rigorous procedures to ensure the veracity of the estimates and proposals contained in the final feasibility study reports. These include:

- Satisfaction that the forecast free cash flows will be sufficient to repay the principal loan amount plus interest on a regular basis, while simultaneously allowing for a margin of safety in those projections to provide the operation with some flexibility.
- Critical appraisal of the project's financial evaluation, with special emphasis on cost and revenue estimates and discount rate used in calculating the project's NPV.

- Technical assessments by expert staff or reputable independent consultants of ore reserve and grade estimates, mine plans and plant designs and operating costs.
- Expert consultants' review of the capital cost estimates to develop or expand the project.
- Examination of the project's management and technical staff.
- Satisfactory commissioning of plant and achievement of pre-defined mine and plant performance.
- Ensuring all appropriate insurances are in place and that the project company has opened all the necessary banking and trading accounts, satisfactory to the bank.
- Regular monitoring of work and expenditure during the developmental and construction stage.

Minimising risk

As stated earlier, forecasts of future business activities can only be an approximation and it behoves those responsible for financial analyses of proposed commercial ventures to ensure their estimates and assumptions are as realistic as possible. There are three variables of particular significance in any financial evaluation of a proposed new mine: commodity sales, operational costs and investment costs. All three influence the cash flows which are the basis of any NPV study.

Examining these factors further we see that commodity sales are dependent on:

- Sale contracts, forward sales and spot market prices
- Prevailing exchange rates
- Production rates
- Mineral or ore grades
- Plant recoveries and efficiencies
- Overall commodity demand.

Estimates of operational costs, on the other hand, can only be maintained if:

- Assumptions regarding the geology of the deposit are correct.
- Mineral/ore grade estimates are accurate.
- Selected mining systems operate in accordance with expectations.
- Mineral processing and any other metallurgical plant operate in accordance with design specifications.
- Estimates of mine and processing plant recovery factors are realistic.

The project's investment costs include all pre-production costs, which embrace:

- Site preparation
- Civil works
- Water and power services and reticulation
- Equipment purchases
- Mine development
- Construction and installation of all fixed plant and equipment.

It follows, therefore, that the quality of the geological, engineering and technological decisions has a great impact on the viability of the future mine. Poor estimation of ore reserves, in terms of both tonnages and grade, would seriously affect the quantity and quality of the saleable mineral commodity and reduce future cash flows. Similarly, if the selected development and production systems do not live up to expectations or if the processing plant performs below par, production levels will be less than expected and once again cash flows will suffer. It is, therefore, wise to have at least one second opinion on these issues before finalising the feasibility study.

Mitigating risk

The previous section highlighted activities that should be implemented to minimise risk. These activities will be incorporated into mine plans at either a feasibility study level or at an operating plan level.

It is more difficult to forecast potential risks associated with markets including commodity price volatility, exchange rate movements, commodity supply and demand fundamentals, government policies impacting resources and other lesser economic risks. However, there is a continuous opportunity for commodity producers to mitigate (reduce or remove) some of these risks through mechanisms including:

- Offtake agreements (mitigates demand risk)
- Commodity hedging (mitigates spot price risk as well as demand risk, considering contango or backwardation contracts)
- Forex hedging (mitigates exchange rate fluctuations)
- Derivative markets (mitigates components of financial risk using market derived instruments including swaps, forwards, options (calls and puts), futures contracts, etc.).

Other risk-mitigating opportunities exist for commodity producers, such as joint ventures, earn-in or farm-in arrangements and securing a cornerstone investor (passive or active), but these and the above-listed financial risk-mitigating activities will not be discussed in explicit detail in this publication. For further information see Lilford (2010).

Discount rates

In discussing cash flows and NPVs of mining projects, the importance of the discount rate used in the financial evaluation cannot be over-emphasised. Discount rates have a near-perfect correlation to risk, and quite often an artificially high discount rate is used to take account of the riskiness of a project. A discount rate is effectively the company's or asset's cost of capital or the opportunity cost of providing capital to the company or asset (Belli et al., 2001).

The difference between the required rate of return on a risky project and that required on a riskless investment is referred to as the risk premium. The discount rate adopted for a risky project is known as the risk-adjusted discount rate.

According to Sorentino and Barnett (1994), and focusing on a specific asset, it is incorrect to assume that the more marginal the asset, the higher the applicable discount rate should be. This is stated on the basis that operating risks or pure project risks should not be factored into a discount rate. Operating risks or the risk associated with not achieving a forecast plan can be factored into the asset's or company's value through its cash flows using other valuation tools. These risk-mitigating tools include:

- Sensitivity analyses incorporating weightings on probable outcomes
- Real options
- Binomial and polynomial tree analysis
- Monte Carlo simulations (Hull, 1997)
- Other available option pricing valuation methods, such as Black-Scholes's theories and formulae (Hull, 1997).

The risks associated with mining projects can range from political and legislative risks that vary from country to country, to economic and financial risks caused by changes in the value of the currency. The risks may also vary according to the mineral commodity as, by way of an example, coal and iron ore projects with long-term sales contracts reflect greater price certainty than base metal projects. Consequently, a company will adopt different discount rates for different commodities due to respective volatilities as well as for different locations due to applicable sovereign risk.

Treasury bonds are considered riskless, and the yield to maturity of government bonds, with their maturity term approximating the duration of cash flow forecasts, is often used as the risk-free rate in the calculation of discount rates. These bond yields vary over time in a similar way in which dividend yields vary. However, they typically range, on a pre-tax basis, between 1.5% and 3.0% (www.bloomberg.com/markets/rates-bonds/government-bonds/australia).

Risk means that a decision could lead to a number of possible outcomes, none of which can be ruled out. Statistics can, however, assist us in assessing and analysing such risks. For example, the standard deviation of Standard and Poor's Index is about 20%. This is a composite index of 500 stocks and, as expected, the standard deviation of a single stock is much higher than that for the market as a whole and is of the order of 35% to 45%. It follows, therefore, that one means of reducing risk is to have a diversified portfolio, albeit that such a portfolio will not totally eliminate market risk but will minimise it and reduce financial returns by using a spread of investments.

The risk associated with a single stock can be measured by how sensitive it is to market movements (Krefetz, 1992). This sensitivity can be assessed by the stock's beta (β) value (Figure 2.1).

The β value of a proposed business venture is a measure of its expected covariance relative to the market as a whole, i.e.

$$\beta = \frac{\text{covariance}\,(r_m, r_s)}{\text{variance}\,(r_m)}$$

The covariance (ζ_{r_m, r_s}) of the r_m and r_s values in a data set is the average product of the deviations about their means.

$$\zeta_{r_m, r_s} = \frac{1}{\pi}\sum (r_m - \bar{r}_m)(r_s - \bar{r}_s)$$

Since r_m and r_s are percentages, this term can also be referred to as the coefficient of linear correlation of these variables.

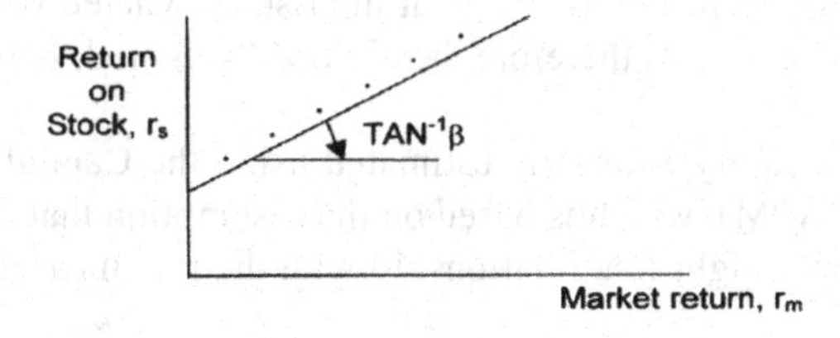

Figure 2.1 Determination of a stock's beta

The variance of a set of numbers is the average of their squared deviations about the mean, such that:

$$\varsigma_{r_m}^2 = \frac{1}{\pi}\sum(r_m - \bar{r_m})^2$$

β is the gradient of the line defining the relationship between r_m and r_s.

If β = 1 any change in r_m results in the same change in r_s
β > 1 any change in r_m results in a greater change in r_s
β < 1 any change in r_m results in a greater change in r_s.

The beta of a portfolio is the weighted average of the betas of the stocks included in it. Thus, if 20% of the stocks have betas of 0.8, 30% have betas of 1.1, and the remaining 50% have betas of 1.3, the beta of the portfolio is:

$$0.2 \times 0.8 + 0.3 \times 1.1 + 0.5 \times 1.3 = 1.14$$

From Lilford (2006), the limitations around the use of a market-based beta in determining a discount rate include:

- Betas indicate the volatility of a specific share price and not of a specific asset within a listed company, such as a mineral property or specific mining operation.
- Betas of a specific listed entity vary as the market varies and not independently of the market.
- Betas vary over time so that the value of a project will also vary over time through a changing discount rate.
- Owing to the cyclicality of the different minerals' prices (supply and demand balances vary differently for each mineral), relative betas will vary so that a perfect correlation becomes improbable.

The value of a company's capital assets is made up of its shares (equity) and borrowings (debt). It follows that the risk associated with a company's capital assets must, therefore, be divided (weighted) between equity and debt.

The cost of equity is usually estimated using the Capital Asset Pricing Model (CAPM), which is based on the assumption that the expected return bears a straight line relationship with the risk measured in terms of its beta.

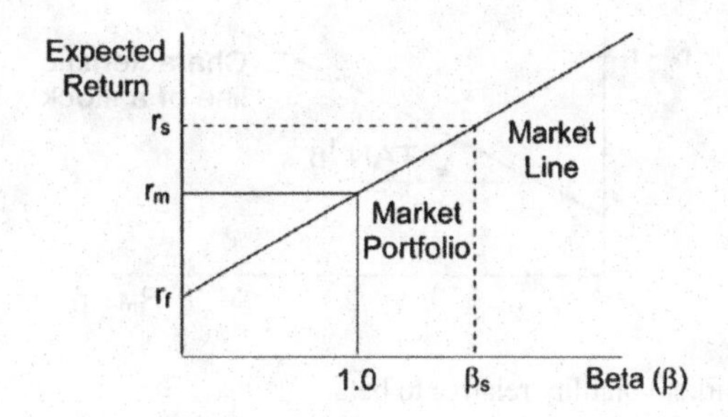

Figure 2.2 Cost of equity relative to beta

Whilst the CAPM is a theoretical model, its relevance has been confirmed using empirical data. It states that the expected risk premium for every stock is proportional to its beta and all assets must be on the market line. Consequently:

Expected risk premium for a stock, S = β_s × expected risk premium for the market
i.e. $(r_s - r_t) = \beta_s (r_m - r_t)$ (see Lilford, 2006; Lilford, 2010 for further details and examples on this)

where r_s = expected return on stock
r_t = risk-free rate
r_m = market rate
β_s = beta value for stock.

The beta value for a stock can be determined from its characteristic line which defines how the stock responds to changes in the market (volatility).

The beta of a company's stock can be estimated by plotting monthly or other regular periodic returns of the stock against monthly or other regular periodic returns on the market index. The slope of the line representing these plots will be the beta of the stock, which demonstrates to what extent that stock's returns correlate with the market returns as a whole.

If we are considering a prospective mining project, its beta value can only be estimated by examining the beta values of operating companies with similar characteristics in the same commodity sector within the same jurisdiction.

So far the betas referred to measure the risk, or volatility, of a company's shares or equity. However, the company's assets will in all probability be

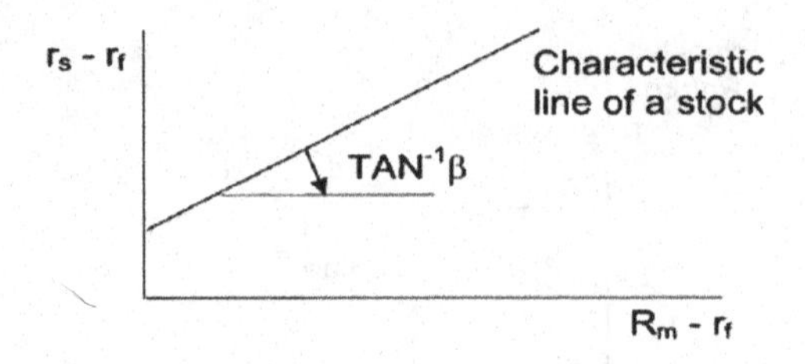

Figure 2.3 Market volatility relative to beta

financed by a combination of debt and equity. The beta of the assets must, therefore, be the weighted average of the betas of both sources of funds, being equity and debt.

$$\beta_{\text{asset}} = \beta_{\text{equity}} \times \frac{\text{equity}}{\text{debt} + \text{equity}} + \beta_{\text{debt}} \times \frac{\text{debt}}{\text{debt} + \text{equity}}$$

The beta of debt has in the past been assumed to be zero so that a company with high beta shares and a high level of debt will have a low beta asset and, consequently, a low cost of capital. For example, if the equity to debt ratio is 1:1 (50% debt and 50% equity funded) and the equity beta is 1.5, we have:

$$\beta_{\text{asset}} = 1.5 \times ½ = 0.75$$

It is also important to point out that the risk associated with a stock, or stocks, in an industry sector has two components, the so called unique risk associated with the company or industry group and the symptomatic or market risk which relates to the market as a whole.

Returning to the CAPM which provides a means of estimating the level of return shareholders expect from their investment, r_s, is given by:

$$r_s = r_f + (\beta r_m - r_f)$$

where r_s = expected return to shareholders
r_f = risk-free rate provided by Government bonds
r_m = market return.

Historically the premium provided by the equity markets over the risk-free rate is between 4% and 6% with the higher figure commonly used. If the post-tax β of a stock is 1.26 and the post-tax return on 10-year government bonds is 5.4%, the post-tax cost of equity capital is given by:

$$r_s = 5.4\% + 1.26 \times 6.0\% = 13.0\%$$

This defines the equity return expected by shareholders on their investment in order to take account of the company's 26% greater share price movement over the market.

Weighted average cost of capital

If a mining company can source all of its funds from the equity market, its cost of capital will be the equity cost of capital estimated by the CAPM. This is, however, rarely the case since most companies also obtain capital from the debt market, because debt is generally cheaper than equity. In this case, it is necessary to obtain the Weighted Average Cost of Capital (WACC) (Van Horne, 1977; Hull, 1989), given by:

$$\text{WACC} = \frac{E}{D+E} \times r_e + \frac{D}{D+E} \times r_d \times (1-t)$$

Where $\frac{E}{D+E}$ = proportion of equity

$\frac{D}{D+E}$ = proportion of debt

r_e = Cost of equity (CAPM)

r_d = Cost of debt

t = corporate tax rate.

Example: Given that there are 100 million shares on issue in a mining company at $2.50 per share and the debt level of project finance is $75 million, calculate the WACC if r_e is 13%, r_d is 9% and t is 30%.

Solution:

$$\text{WACC} = \frac{250}{250+75} \times 13\% + \frac{75}{250+75} \times 9\% \times 0.70$$
$$= 11.45\%$$

It should be borne in mind that the WACC in this case is an after-tax calculation and should therefore only be applied to after-tax cash flows. It is also important to note that WACC discount rates take the risks associated with the future volatility of commodity prices into account. Consequently, if a hedging program is in place to fix the future price of the saleable mineral commodity, or a proportion of it, the WACC discount rate should ideally be adjusted to take account of this fact.

The WACC discount rate is also the nominal post-tax cost of capital which assumes that inflation is present. It follows, therefore, that if this discount rate is to be used in a DCF analysis, all the cash flows should be current or nominal values. Rather than estimate future inflation rates, it is preferable to adjust the discount rate to produce a constant or real discount rate using the Fisher equation (Fisher, 1977):

$$1 + \text{real rate} = \frac{(1 + \text{nominal rate})}{(1 + \text{inflation rate})}$$

Referring to the previous example and assuming an annual inflation rate of 2.5% we have:

$$1 + \text{real rate} = \frac{(1 + 0.1145)}{(1 + 0.025)} = 1.0873$$

$$\text{real rate} = 8.7\%$$

Geared and ungeared betas

As stated earlier, the beta factor measures the risk or volatility of an investment relative to a well-diversified portfolio. The betas of companies derived from stock market data will incorporate both equity and debt financing. It is, therefore, impossible to compare the equity betas of companies without taking into account their gearing levels. The true equity betas or ungeared betas, explained in detail in Lilford (2010), of companies can be determined as follows:

$$\beta_u = \frac{\beta_g}{\left[1 + \left(\frac{D}{E} \times (1 - t)\right)\right]}$$

Where β_u = ungeared beta
β_g = geared beta
D/E = debt to equity ratio
t = corporate tax rate.

Example: Given that the geared beta of a company is 0.95 and its net debt to equity ratio is 25%, calculate its ungeared beta.

Solution:

$$\beta_u = \frac{0.95}{[1 + .25 \times 0.7]}$$
$$= 0.8$$

(Note: Net debt is interest bearing debt less cash.)

Carrying out such calculations for a representative cross-section of an industry sector (e.g. gold mining sector) will provide a mean value of ungeared beta for those companies included in the study. This mean value will, however, have to be regeared if it is to be applied to a new project or a company under investigation. If, for example, the company being valued has a 20% debt to equity ratio and the mean ungeared beta for a representative group of companies is 0.8, the regeared beta is given by:

$$\beta_g = \beta_u \times \left[1 + \frac{D}{E} \times (1 - t)\right]$$
$$= 0.9 \times 1.14$$
$$= 1.03$$

Varying discount rates

Since a discount rate reflects a point in time which in turn accommodates a specific weighting in the cost of equity and the cost of debt, it is logical then that as the debt to equity ratio changes, so should the discount rate. Based on the previous discussions determining the WACC and CAPM results which determine a discount rate, Table 2.3 highlights the impacts that specific periods will have on the discount rate as the debt to equity ratio changes.

The above tabled example clearly shows that using a single discount rate over the life of the asset may be a flawed approach and that varying

Table 2.3 WACC changes as debt: equity ratio changes

		Start	*Peak*	*Steady*
Debt Level	$'m	50.0	90.0	30.0
Equity Level	$'m	80.0	80.0	80.0
Debt/Capital	%	38.46%	52.94%	27.27%
Tax Rate	%	0.00%	30.00%	30.00%
Borrowing Cost (pre-tax = 6.50%)	%	4.550%	4.550%	4.550%
Unlevered Beta		0.85	0.85	0.85
Equity Risk Premium	%	5.00%	5.00%	5.00%
Risk-Free Rate	%	1.68%	1.68%	1.68%
Re-levered Beta		1.38	1.52	1.07
Cost of Equity	%	8.59%	9.28%	7.05%
Proportional Cost of Equity	%	5.28%	4.37%	5.12%
Proportional Cost of Debt	%	1.75%	2.41%	1.24%
WACC nominal	**%**	**7.03%**	**6.77%**	**6.37%**
WACC real (3.5 CPI)	**%**	**3.41%**	**3.16%**	**2.77%**

the discount rate over time should be considered. The period-by-period changes to the company's or asset's risk structures should be considered and given an appropriate weighting in the long-term cash flows (Laughton and Jacoby, 1991). In summary, a constant discount rate may bias project alternatives.

Finally, looking at pure project risk, some literature suggests and recommends the incorporation of project risk into discount rates (Davis, 1995), while others consider this practice incorrect (Sorentino, 1993). Whichever way is argued, each side of the argument carries weight. However, the cost of capital and the cost of equity must not be confused with project risk. To ensure that the discount rate reflects risks associated with funding and sovereign aspects, it is simple to apply a differential discount rate to determine the impacts of each risk class separately. In Table 2.4, it is assumed that the discount rate has been determined to be 10% over ten years with a pure project risk discount varied over the first few years, settling at 1% over the final eight years.

Intuitively from the table, it is simple to adjust the discount rate on a periodic (annual) basis to compensate for varying debt : equity ratios over time as well as to accommodate project risk in a separate risk-factor without incorporating it into the discount rate.

Table 2.4 Separating the discount rate from project associated risk discounting

Discounting (Years)		*1*	*2*	*3*	*4*	*5*	*6*	*7*	*8*	*9*	*10*
Discount rate	%	10%	10.0%	10.0%	10.0%	10.0%	10.0%	10.0%	10.0%	10.0%	10.0%
Cash flow	$'000	−79,769	−24,538	19.103	40,924	40,924	38,275	26,600	26,600	26,600	19,950
Discount factor (rate)		0.909	0.826	0.751	0.683	0.621	0.564	0.513	0.467	0.424	0.386
Discounted cash flow	$000	−72,517	20,280	14,352	27,951	25,410	21,605	13,650	12,409	11,281	7,692
Progressive discounted value	**$'000**	**−72,517**	**−92,797**	**−78,444**	**−50,493**	**−25,083**	**−3,477**	**10,173**	**22,582**	**33,863**	**41,555**
Pure Project RiskDR	%	4%	2.0%	1.0%	1.0%	1.0%	1.0%	1.0%	1.0%	1.0%	1.0%
Project Risk Discount Rate		0.962	0.943	0.933	0.924	0.915	0.906	0.897	0.888	0.879	0.871
Risk-adjusted value	$000	−69,728	−19,117	13,396	25,830	23,249	19,572	12,243	11,020	9,919	6,696
Progressive Value (risk-adjusted)	**$'000**	**−69,728**	**−88,845**	**−75,450**	**−49,619**	**−26,370**	**−6,798**	**5,446**	**16,466**	**26,385**	**33,081**
Progressive IRR on Cash flow	%			−64.08%	−20.20%	−1.09%	8.77%	13.15%	16.19%	18.33%	19.51%

Source: Lilford (2017).

References

Belli, P., Anderson, J.R., Barnum, H.N., Dixon, J.A. and Tan, J.P. 2001. *Economic Analysis of Investment Operations: Analytical Tools and Practical Applications*. Washington, DC: World Bank Institute Development Series.

Davis, G.A. 1995. *(Mis)use of Monte Carlo Simulations in NPV Analysis*. SME Non Meeting Paper 94–309. Manuscript April 11, 1994.

Fisher, I. 1977. *The Theory of Interest*. Philadelphia: Porcupine Press.

Hull, J.C. 1989. *Options, Futures, and Other Derivatives*. Third Edition. Upper Saddle River, NJ: Prentice-Hall.

Hull, J.C. 1997. *Options, Futures and Other Derivatives*. Upper Saddle River, NJ: Prentice-Hall.

Krefetz, G. 1992. *The Basic of Stocks*. Chicago, IL: Dearborn Financial Publishing Inc.

Laughton, D.G. and Jacoby, H.D. 1991. *Project Valuation with Lognormal Reversion in Prices*. University of Alberta Institute for Financial Research. Working Papers, pp. 3–91.

Lilford, E.V. 2006. The Corporate Cost of Capital. *The Journal of the South African Institute of Mining and Metallurgy*, 106, February 2006.

Lilford, E.V. 2010. *Advanced Methodologies for Mineral Project Valuation*. Australian Institute of Geoscientists. ISBN: 1 876118 41 5. ISSN: 0812 60 89.

Lilford, E.V. 2017. *Lecture Notes, Resources Sector Management, Department of Mineral and Energy Economics*. Curtin University.

Sorentino, C. 1993. *Mine and Property Evaluation: Risk Analysis*. Notes for a Graduate Course. Master of Minerals and Energy Economics. Macquarie University.

Sorentino, C. and Barnett, D.W. 1994. *Financial Risk and Probability Analysis in Mineral Valuation*. Proceedings of VALMIN 94, The Australasian Institute of Mining and Metallurgy, Carlton, Australia, ISBN 1 875776 036, pp 81–101.

Van Horne, J. 1977. *Financial Management and Policy*, Fourth Edition. Upper Saddle River, NJ: Prentice-Hall, pp. 84–95, 197–225.

3 Royalty pricing, agreements and contract systems in mining

Introduction

It is important to understand that a royalty is not explicitly a tax. It is actually a payment to a property owner in exchange for the right to use that property, or in this case exploit a mineral deposit or orebody. If, as is the case in Australia and many other mineral-rich countries, the minerals in the ground are owned by the government on behalf of the people; then it is considered appropriate that mine owners should recompense the government for the exploitation of these non-renewable mineral assets as a payment beyond or in addition to taxes on profits. This may manifest as a rent payment or as a royalty payment.

For clarity purposes, to introduce a definition of ore, which is the material mined for the purpose of extracting minerals or metals and which may be sold untreated (ex-headframe) for its content of mineral(s) or metal(s) (Harries, 1996). On the other hand, the mined material may be processed to remove the mineral(s) or metal(s) before selling the mineral concentrate (ex-mill) or metal(s). Ore does not include waste material, even though it may be used for such applications as backfilling, rail ballast or road building.

In most mineral-rich countries, at least one specific government department will be responsible for:

- Developing and providing royalty policy advice.
- Implementing the royalty policy, if applicable.
- Assessing, collecting and verifying royalties.
- Preparing and monitoring royalty budgets and forecasts.

As can be expected, there are a number of royalty payment types used in different countries, and there are hybrid structures incorporating aspects of these royalty types. As detailed by Boadway and Keen (2014), there are five main royalty payment types that exist as follows:

- Specific or unit-based rate: flat rate per tonne or volume based;
- Ad-valorem: percentage of value;
- Profit-based: percentage of profit;
- Economic rent based: percent above break-even;
- Production sharing contracts: percent of product.

The specific or unit-based rate royalty system is the easiest to police and administer, but it takes no account of varying ore grades or cost factors which vary from mine to mine. Consequently, this system is largely limited to low value construction materials or to low unit value bulk commodities.

The ad-valorem or value-based royalty system is calculated as a percentage of the value of the mineral, typically being the sales value of the mineral or the income derived from the sales of that mineral recognising minimal deductions. Except for gold, this means the gross invoice value of the mineral (less any allowable deductions for the mineral, as defined in the Mining Regulations and excludes on-site operating costs). That is, the allowable deductions relate only to the costs incurred in transporting the mineral product overseas or packaging the material being readied for export or transport in the country of origin.

The rate of an ad-valorem royalty around the world ranges up to, typically, 10%. As an example, the Democratic Republic of Congo uses a rate of 10% for cobalt sales and 3.5% for copper sales, although it is contemplating designating copper as a strategic mineral and hence increasing its royalty rate to 10% too to match that of cobalt. Generally, the royalty rate will reduce in the event that the commodity is upgraded to near-metal or even further if to a refined metal.

Profit-based royalties are calculated as a percentage of net profit. The profit is calculated by deducting allowable project costs, including all operating costs, from project revenues. The benefit to the operator or owner of the project with this profit-based royalty is that if the operation does not make a profit, no royalty is paid, unlike the ad-valorem royalty which does not consider the profitability of the operation at all.

Royalty systems between private parties

Royalties refer to payments which may be made to the State or which may be negotiated between private individuals for the right to mine an orebody, lease a mining property and/or gain access to a property. Royalty payments may also be the means by which a joint venture partner ceases to be an active partner in a mining venture with that partner selling his joint venture equity holding to the other existing partner in exchange for a royalty.

Harries (1996) has published an excellent review of royalty agreements between private parties and, although it is written for the Canadian and North American markets, the subject matter has universal applicability.

In Canada, the early royalty systems for precious metals were usually 'net smelter return' or 'gross' royalty systems with rates of up to 5%. However, as the cost of mining increased with increasing orebody depths and the mining of base metals or other minerals involving complex and expensive metallurgical processes, such rates became untenable. As a consequence, the net profit royalty or net profit interest system came into use (Harries, 1996), since the investor or operating company had the property for nothing unless and until it proved to be profitable. In this case higher royalty rates applied, but unfortunately such systems were, and still are, open to abuse. Despite this, a variant of this being the Net Smelter Return (NSR) system is often preferred, even though the rates are lower (1% to 3%) to take into account higher operating and treatment costs.

A royalty agreed to between two private parties does not obviate the operation from paying a government royalty, if applicable.

Net smelter return royalty

Net smelter return royalty refers to a percentage payment based on the gross revenue from the disposal of the mineral product less a few deductions, such as the cost of shipping the product to the purchaser, sampling and assaying, insurance charges and direct taxes and imposts. The term NSR is in many ways a misnomer since, as Harries (1996) states, in most cases NSR royalties are paid for products that have never seen the glow of a smelter. It is also the case that the alternative term of 'gross royalty' is misleading since, explicitly, it means payment without any deductions, and whilst such agreements exist, the term is often used as an alternative to NSR which does allow for a few specific deductions.

NSRs usually refer to the funds received from the sale of ores or from the sale of concentrates or other products derived from ores, less the treatment costs and all costs associated with insurance, freight, trucking, handling and/or sampling and assaying the ores, concentrates or other products. These are typically measured exheadframe for ores and ex-mill or other treatment facility for concentrates or other products. In addition, government taxes and costs and expenses associated with custom milling, smelting, refining or other treatment of ores, minerals or metals are also deductible before calculating the NSR.

As an example, in Canada, the NSR is calculated by multiplying the quantity of gold, silver or other metal or minerals contained in the product and sold or disposed of during any one year to a purchaser by an agreed benchmark price for that commodity. The contracted benchmark price will be an acceptable, internationally recognised and undisputable price, or average price, quoted by an acceptable trading business or possibly by an investment or merchant bank. Without allowing any deductions, this would give rise to a Gross Smelter Return (GSR), whereas an NSR equates to a GSR less the total of all allowable deductible amounts (smelting and refining costs and costs associated with transport including various insurance costs) paid or incurred during that year. If the saleable product has been hedged (sold into the forward market at predetermined prices), the NSR will typically recognise the hedged prices as being the price at which the product is sold. Since less than 100 percent of any product is sold into a forward contract in any one production year, the realised price for the NSR calculation will be the weighted price of the hedged price and the market price as listed above.

Elucidating on the allowable deductions for an NSR, the deductible amounts refer to all costs or expenses paid or incurred by or for the account of the producer during the year with respect to:

- Costs associated with insurance, security, freight, trucking, handling and/or weighing, sampling and assaying of the ex-headframe ore or the ex-mill concentrate or the product from other treatment facilities.
- All costs, charges and deductions incurred in marketing or selling the product.
- Any and all deductions, charges or penalties applied by the purchaser due to the chemical composition or physical attribute of the product.
- Any federal, state or municipal tax or levy relating to the product.

It is important to bear in mind that a mine that originally planned to produce gold or some other metal or mineral may turn out to be a mine endowed with other economically recoverable metals or minerals. It follows, therefore, that royalty agreements should be written in such a way that all possible products are included.

It is also worth emphasising that the NSR for a gold producer is calculated on the basis of the average of the second daily fixes for gold at the London Metal Exchange for each month and the average of the prices for other recovered and sold metals quoted each month in mutually acceptable price-source publications, multiplied by the quantity of gold or other recovered and sold metal in the product.

Net profit royalty system

Whereas the NSR system is fairly straight forward and easy to understand, the Net Profit Royalty (NPR) system tends to be both complex and difficult to comprehend. It is also administratively more difficult and onerous to manage. It usually requires a great deal of information and the services of a qualified accountant. It is also much more open to abuse and has led some cynics to refer to the acronym for the Net Profit Interest (NPI) system as 'no profits intended'.

On the other hand an NPR or NPI system can, if determined equitably, be fair to and supported by both parties. However, one of its disadvantages is that the recipient has to wait for a return, since the recipient agrees to share in the risks associated with the venture. Consequently, if there is no profit there is no royalty. Another problem for the recipient is the need to use the services of expert professionals to ensure the accuracy of all the entries and calculations required in the calculation of the royalty payments.

As Harries (1996) states, NPR system should not be discarded out of hand, but there is no doubt it does require careful structuring and control. Harries's definition of the terminology associated with the system is as follows:

> ***Net Profit*** is usually calculated annually as the aggregate of revenues received from the mining, milling and/or other treatment of any ores or concentrates and/or marketing of any product, including cash proceeds received from the sale of capital assets in the ordinary course of the business or in anticipation of the termination of the business or from the investment of moneys retained with respect to such operations, less:
>
> - All or part of the amount by which operating costs for any prior year(s) exceeded revenues.
> - The aggregate of all operating costs incurred during the year.
> - The aggregate of all preproduction expenditures incurred by the mine owner.
> - Such amount as may be required to maintain working capital at an amount considered by the operator to be advisable in order to carry out operations on the property in a proper and efficient manner.
> - Reserves for contingencies which are confirmed by auditors.
> - The costs of any major improvement, expansion, modernisation and/or replacement of the mine, mill or ancillary facilities.
>
> ***Operating Costs*** for any year is the amount of all expenditures or costs incurred in connection with the business of mining, milling and/or

other treatment of ores or concentrates, and/or marketing any product resulting from the operations at the time including all costs:

- relating to:
  - mining, crushing, handling, concentrating, smelting, refining or other treatment of ores or concentrates;
  - the handling, treatment, storage or disposal of any waste materials and/or tailings; and
  - the operation, maintenance and/or repair of any mining, milling, handling, treatment, storage or ancillary facilities related to the business;
- of, or related to, marketing any product including transportation, commissions and/or discounts;
- of maintaining the property in good standing or renewing permits from time to time;
- of providing operating facilities and/or accommodation;
- all duties, charges, levies, royalties, taxes and other payments imposed on the business by government or municipality;
- and fees payable for providing technical, management and/or supervisory services;
- of bringing the mine into commercial production, including the payment of interest and/or other fees;
- associated with consulting, legal, accounting, insurance and security services;
- of construction, equipment and mine development after the commencement of production, including maintenance, refurbishments, repairs and replacements;
- of pollution control and mine closure requirements associated with government regulations;
- and any royalties or similar payments paid to any third party; and
- and any expenses incurred as a result of terminating the business, including disposal of assets, termination of employees and mine rehabilitation.

Preproduction Expenditures are the total of all costs, capital or otherwise, relating to exploration and development of the mine and bringing it into commercial production and/or construction of facilities relating thereto, including all costs:

- and monies spent in doing work until the property has been brought into commercial production;

- associated with the construction of the mine, mill buildings, crushing, grinding, washing, concentrating, waste storage and/or disposal facilities;
- of exposing and mining the orebody or orebodies, but only until the commencement of production;
- of constructing storage and/or warehouse facilities, constructing and/or relocating roads and the acquisition and/or developing waste and/or tailings disposal systems;
- of transportation facilities for moving ore, concentrates and/or any products derived therefrom, electric power and power lines, equipment, water pipelines, pumps, dams and wells;
- of employee facilities;
- of supplying management, marketing, supervisory, engineering, accounting or other technical and/or consulting services or personnel;
- of leasing and/or maintaining the property in good condition;
- of feasibility, marketing, economic, reclamation, rehabilitation, and/or technical evaluations, plans, studies or reports;
- relating to consulting, legal and marketing activities up until commencement of production; and
- associated with the financial arrangements relating to bringing the mine to commercial production, including the payment of interest and/or fees or charges until commencement of production.

Net Proceeds are calculated by deducting all costs and expenses from the gross revenues, which are the total monies received from the sale of mineral commodities. Costs and expenses are those amounts associated with the exploration, development, construction, exploitation and marketing of the minerals, and any net loss accrued during an accounting period is carried forward to reduce the net proceeds in ensuing periods.

Royalties paid to governments

The Bradley Report (1986) reviewed the manner by which royalties can be raised by State governments in Australia from producers of mineral or fossil fuel resources. As will be seen, the terminology used to describe royalty systems can vary, but in general there are three main systems:

- The gross or quantum royalty system, based on a fixed charge per unit of production.
- The ad valorem royalty system, which is a duty levied as a percentage of the value of the mineral or fossil fuel produced.
- The profit-based royalty system, which is usually applied as a fixed percentage of the profit, defined in an appropriate way.

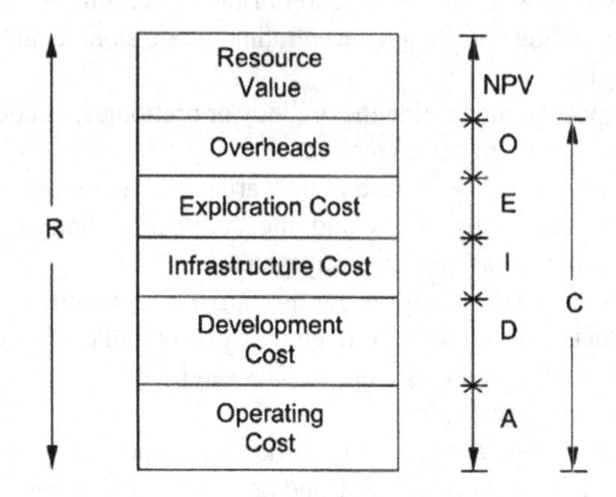

Figure 3.1 Value components of a mineral deposit

Mineral and fossil fuel resources in Australia and many other jurisdictions are owned by the states or government. It is, therefore, logical for the states or government to claim a financial return when the title to a particular resource is transferred to the developer. This financial return is obtained as a royalty, and its magnitude depends on the value of the resource.

The licensing/leasing system associated with the development of mineral resources proceeds through stages. It usually commences with a prospecting or exploration license and may later be upgraded to a mining lease. Application and periodic rental fees are charged for these licenses and leases to meet the costs of administration.

Royalties, however, usually occur only when production commences.

Following are the elements which make up the value of a mineral deposit or of an oil and/or gas reservoir (Figure 3.1).

In this schematic diagram, R represents the revenue generated by the sale of a commodity measured in present value dollars. The total cost, C, comprises many components and is also measured in present value dollars. The resource value is the net present value (NPV) of the deposit.

It therefore follows that:

$$NPV = R - C$$

and $C = A + D + I + E + O$

where A = Estimated present value of the direct operating costs incurred during the production phase.

D = Estimated development investment in present value dollars, which includes site preparation, decline development and/or shaft sinking, etc., and plant and equipment costs.

I = Estimated infrastructure expenditure involved in establishing essential services such as road and rail access, township facilities, etc., required to sustain production.

E = The costs associated with exploration to define the economic attractiveness of possible exploitation.

O = This element includes fixed costs such as rent and leasing charges and also includes an allocation to take account of exploration undertaken elsewhere which the company must engage in to remain in business.

Alternative royalty systems

There are a variety of royalty systems in use, including gross royalty systems and those based on accounting or economic profits. Gross royalty systems involve either a specific charge on production as dollars or cents per tonne or a fixed percentage charge based on the value of production. The concept of accounting profits is used by some in assessing corporate income tax. It allows for the deduction of most costs before calculating profits and the depreciation of capital expenditure according to clearly defined rules. Interest charges are also deductible, but the return on shareholders' investment is not.

The concept of economic profit, on the other hand, allows for all costs, including normal returns to shareholders, to be taken into account. Whilst there are also many hybrid systems, the following overview covers the main systems.

All royalty systems impose an additional cost on mineral production and consequently reduce the net value of all deposits. They can therefore reduce the attractiveness of marginal projects and/or those parts of the deposit which involve higher extraction costs.

Gross Royalty – Specific Duty (GRSD)

This is a duty levied on the volume of mineral production and is charged as dollars or cents per tonne if the ore is of a more or less constant grade, or dollars or cents per pound of contained mineral if there is significant variation in the ore grade. The system can be modified to take some account of price movements by linking it to an indicator of product value.

The GRSD is basically an attractive system because of its ease of application, the low cost of its administration and the constant level of government revenue it generates.

Gross Royalty – Ad Valorem Duty (GRAVD)

This is a duty levied on the value of mineral production and is, therefore, dependent on both the volume of production and the selling price for the mineral commodity. Whilst it may be set initially with reference to a resource value, it will not subsequently respond to changes in extraction costs. Thus, for a given deposit the royalty paid as a fixed percentage on the value of mineral output will vary in real terms as the costs of production change.

This duty can be levied at any point in the production process, e.g. Free On Board (f.o.b.), Free On Rail (f.o.r.), or ex mine. The further downstream one goes, the larger the value-added component. Hence, a GRAVD levied at a fixed percentage will result in higher monetary values as the point at which the levy is applied moves downstream.

Whilst the GRAVD is fairly easy to apply and administer, it does need additional information about mineral commodity values. Prices may be difficult to define in vertically integrated companies, and separating mining and mineral processing costs may be necessary if the royalty is to be assessed at the mine-head.

The income generated by GRAVD varies with both market price and production volumes and is, therefore, more volatile than with GRSD systems.

Accounting Profits-Based Royalties (APBR)

This system is applied as a percentage of the accounting profit, which is defined as revenue less expenses. Operating costs are fully deductible and capital expenditure is depreciated according to established accounting practice. Interest on borrowed funds is a legitimate cost, whereas normal returns on shareholders' equity is not, and consequently accounting profits are overstated when compared to economic profits. In applying the system, it is usual to adopt the accounting conventions developed for corporate income tax purposes.

Apart from the distorting effects of using accounting profit defined as the excess of revenues over allowable costs of production, depreciation calculated in accordance with tax regulations can also lead to a misleading picture of the cost of owning an asset.

Nevertheless, this system is more equitable than gross royalty systems because it does distinguish between marginal mining operations and those which exploit highly profitable deposits. It also has the advantage of adjusting automatically to changes in prices and costs.

On the other hand, the system is more expensive to set up and administer than those based on gross royalty systems.

Resource Rent Royalty

This system is based on economic profit, which is defined as revenue net of all costs, inclusive of the return expected by equity investors. The method involves government levying a royalty based on a given rate on revenues, less expenditure in each period, with no distinction being made between operating and capital expenditures.

It also follows that Resource Rent Royalty (RRR) systems make no distinction between positive and negative net returns, so that governments would make a payment if the net return for a specific period was negative. Stated differently, this system requires governments to share in the risk of resource development in proportion to their royalty share so that they become suppliers of capital during the capital expenditure phase of projects. For this reason, the RRR system has never been accepted as a viable system.

Nevertheless, a modification of the RRR system was adopted by the Australian Commonwealth Government for greenfields offshore petroleum production. Such modified systems, based on the NPV of the project, ensure that payment is realised only when the cumulative NPV is positive, and it is this value that is subject to the royalty at the set rate. Intuitively, this system involves two variables, being the discount rate (threshold rate) and the royalty rate. In the case of the Commonwealth's resource rent tax referred to earlier, a single threshold rate is applied to all petroleum extraction.

One disadvantage of the modified RRR system is its potential variability, since the royalty is dependent on mineral prices and payments are only made when the project's NPV becomes positive. With gross royalty systems, royalties are forthcoming from the start of production. Its administration would also be more difficult than for an APBR system, because of the need for continuing economic analysis due to the importance of the threshold rate parameter. On the other hand, the RRR does not require the use of depreciation conventions.

Royalty systems in Western Australia

The legal system in Australia ensures that when State governments release land for development, all mineral and petroleum resources are withheld on behalf of the community. That is, the community retains ownership of the natural resources. Royalties are therefore payments made to the State government by those exploiting the community's resources in exchange for the right (license) to do so.

State governments, as custodians of their mineral and petroleum inventories, often determine or must agree to the system and level of royalty payments associated with a mining proposition.

Mineral royalties in Western Australia are collected under the auspices of the Mining Act 1978, the Mining Regulations 1981 or one of the special State Agreement Acts negotiated for individual projects.

The following are the three main systems of royalty collection in use in Western Australia, depending on the degree to which the mineral has been beneficiated before sales:

- Specific rate, being quantity based, is calculated as a flat or predetermined fixed rate per tonne of production (generally used for low-value construction and industrial minerals);
- Ad valorem, calculated as a percentage of the value or revenue derived from the mineral sold, less any allowable deductions; and
- Project based, calculated as a percentage of profit (similar to a normal tax).

Specific rates in current use are typically 73–117 cents per tonne, until 2020 at which point they will be reviewed. While this system is easy to administer, it takes no account of either ore grade or cost factors, which will differ from one mine to another, and is consequently only applied to low value construction minerals. Approximately less than 5% of the state's mineral royalty income comes from this source.

Ad valorem or value-based royalties are determined as a proportion of the gross sales value. It is generally charged as a percentage of the free on board value of minerals to be exported, or alternatively, it is based on the realised value of the free on rail or free on truck value of the product. The state's current ad valorem system is based on the following principles:

- Ore subjected to limited treatment (bulk material) 7.5% of the royalty value;
- Mineral concentrate (significant upgrading through a concentration plant) 5% of the royalty value; and
- Metal, 2.5% of the royalty value.

This system takes account market price fluctuations and varying ore grades, but excludes any and all production costs. Around 85% of the state's income from mineral royalties is generated by ad valorem royalty systems.

Profit-based systems are, as with normal corporate taxes, based on a fixed percentage of the net profit, which is calculated by taking into account all revenues and project costs. Whereas it is acknowledged to be the most equitable system, it has the disadvantage of generating an uncertain revenue and is difficult to administer. Not surprisingly there is, therefore, no profit-based system currently in use in Western Australia.

There is, however, a combined ad valorem/profit-based system in use for the Argyle Diamond Project. This system provides a profit-based royalty of 22.5% over the life of the project, whilst guaranteeing a minimum ad valorem royalty payment of 7.5% in any one year. This mining project alone contributes significantly to the total royalty revenue from the minerals sector in Western Australia.

The state's petroleum royalties are governed by a number of Petroleum Acts or Agreements. The State's Petroleum Act covers onshore projects and those in coastal waters within a line which follows the shoreline and islands. These projects are charged a royalty based on a well-head value system and all such revenues are retained by the State government.

The area covered by the Territorial Sea, which extends three nautical miles from the onshore and coastal waters boundary line, comes under the jurisdiction of the State's Petroleum (Submerged Lands) Act. All projects in this area also pay royalties based on a well-head value system to the State government, albeit that 40% of this income is passed on to the Commonwealth.

The offshore areas extending beyond the Territorial Sea to the boundary of the Australian Continental Shelf are covered by the Commonwealth's Petroleum (Submerged Lands) Royalty Act for projects which were in existence prior to 1 July 1984. This is, therefore, the Act which applies to Woodside's North West Shelf Project. Royalties from these projects are paid to the Commonwealth, although Western Australia receives 68% of the income in order to cover the cost of its administrative functions. Projects developed after 1 July 1984 are subject to the Commonwealth's Petroleum Resource Rent Tax (PRRT), which is a net cash flow royalty system.

The well-head value is the gross sales value less all costs incurred between the well-head and the point of sale. Such costs include processing, storage and transport activities.

Whereas the basic royalty rates range from 10% (primary production license) to 12.5% (secondary production license) of the well-head value, they have been as low as 5%. A 5% royalty was agreed to for the first 15 years of West Australian Petroleum Pty Ltd's (WAPET's) Barrow Island operation, which ended in 1982. This low rate was agreed to in lieu of the company's extensive exploration program.

Commonwealth legislation, in addition to State royalties, also provides for an excise on all oil and condensate produced onshore or within three nautical miles of the Australian coastline, and captures some offshore North West Shelf production in this net.

There are three systems used to capture petroleum royalties, being:

- Well-head royalties, derived by taking the gross value of petroleum produced less all costs incurred beyond the Christmas tree and the sales point. It therefore excludes exploration and recovery costs.
- Resource Rent Royalty (RRR).
- Petroleum Resource Rent Tax (PRRT).

The Commonwealth, State and Territory governments in Australia developed a Hybrid Ad Valorem and Net Income Royalty (HANIR) system for the offshore industry. It is based on a low first-tier value-based royalty as well as a second-tier net income royalty. The 1st tier commences as soon as sales begin and the second tier kicks in once the project becomes profitable, and is calculated as a percentage of net income. It is claimed this system achieves a compromise between the three main objectives of a royalty system, i.e. economic efficiency, revenue stability and administrative simplicity.

Royalty charges for minerals in Western Australia

The payment arrangements and amount of royalties relating to the exploitation of minerals owned by the Crown are outlined in legislation under either the Mining Act, Petroleum Acts or Special Agreement Acts.

The Mining Act 1978 defines minerals as naturally occurring substances including evaporites, limestone, rock, gravel, shale, sand and clay.

The following table and associated notes define the royalty charges for most minerals exploited in Western Australia.

Rates of royalty in respect of gold

- When gold metal is produced from gold bearing material that was produced or obtained from a mining tenement, royalties shall be paid by the holder of, or applicant, for the mining tenement.
- The rate of royalty payable for gold metal produced after 30 June 2000 is 2.5% of the royalty value of the gold metal produced.
- No royalty is payable in respect of the first 2,500 ounces of gold metal produced during a financial year from gold bearing material produced or obtained from the same gold royalty project.
- The royalty value of gold metal produced shall be calculated for each month in the relevant quarter by multiplying the total gold metal produced during that month by the average of the gold spot prices for that month.

(www.dmp.wa.gov.au/Minerals/Royalties-1544.aspx)

Petroleum royalties

Petroleum royalties in Western Australia are administered and collected under State and Commonwealth legislation. Royalties from onshore projects are kept by the State government, whereas royalties collected from offshore projects are shared between the State and Commonwealth in accordance with the relevant legislation. Over half a century ago, State and Commonwealth officials agreed and implemented an Offshore Constitutional Settlement. The agreement captures a 60:40 revenue sharing arrangement

	Column 1	Column 2	Column 3
Mineral	*Amount per tonne according to quantity produced or obtained*	*Percentage of the royalty value*	*The rate as specified hereunder*
Aggregate	Amount A		
Attapulgite		5%	
Bauxite		7½%	
Building Stone	Amount B		
Chromite		5%	
Clays	Amount A		
Coal (including lignite) – not exported			$1 per tonne, to be adjusted each year at 30 June in accordance with the percentage increase in the average ex-mine value of Collie coal for the year ending on that date when compared with the corresponding value of Collie coal for the year ending on 30 June 1981.
– exported		7.5%	
Cobalt		7.5%	
if sold as crushed or screened material –		5%	
if sold as a concentrate –		2.5%	
if sold as a nickel by-product or in metallic form –			
Copper		7.5%	
if sold as crushed or screened material –		5%	
if sold as a concentrate –		2.5%	
if sold as a nickel by-product or in metallic form –			
Diamond		7½%	

(*Continued*)

(Continued)

	Column 1	*Column 2*	*Column 3*
Mineral	*Amount per tonne according to quantity produced or obtained*	*Percentage of the royalty value*	*The rate as specified hereunder*
Dolomite	Amount A		
Feldspar		5%	
Garnet			The rate shall be – (a) 5% for the usual grades of garnet including that used for sand blasting and filtration; (b) 2.5% for higher technology grades including that used for garnet paper and polishing purposes, of the royalty value, calculated on the basis of the nearest available port if exported.
Gems and Precious Stones		7.5%	
Gravel	Amount A		
Gypsum	Amount A		
Ilmenite (other than ilmenite feedstock as defined in regulation 86AC)		5%	
Iron Ore (including magnetite) –			
beneficiated ore (iron ore that has been concentrated or upgraded otherwise than by crushing, screening, separating by hydrocycloning or a similar technology, washing, scrubbing, trommelling or drying, or by a combination of two or more of those processes)		5%	

iron ore other than beneficiated ore		7.5%	
Kaolin		5%	
Lead			The rate is – (a) if sold as a concentrate, 5% of the royalty value; or (b) if sold in metallic form, 2.5% of the royalty value.
Leucoxene		5%	
Limestone (including limesands and shellsands) –			
used for agricultural or construction purposes or as a neutralising agent in tailings treatment operations	Amount A		
used for metallurgical purposes (other than as a neutralising agent in tailings treatment operations)	Amount B		
Lithium Minerals		5%	
Manganese		7.5%	
Manganese (beneficiated by the producer in Western Australia otherwise than by crushing, screening, washing, scrubbing, trommelling or drying, or by a combination of two or more of those processes)		5%	
Nickel		2.5%	
Ochre		5%	
Platinoids		2.5%	

(Continued)

(Continued)

	Column 1	*Column 2*	*Column 3*
Mineral	*Amount per tonne according to quantity produced or obtained*	*Percentage of the royalty value*	*The rate as specified hereunder*
Rare earth elements			In accordance with the following formula: $\frac{p}{100} \times \frac{U2.5}{100} = \$R \text{ per tonne}$ Where p = a representative market value of rare earth oxides (REO), as determined from time to time by the minister. Where U = the number of units per hundred of REO in the rare earth elements-containing products sold. Where R = the royalty.
Rock	Amount A		
Rutile		5%	
Salt	Amount A		
Sand	Amount A		
Semi-precious stones (including specimen stones)		7.5%	
Silica	Amount B		
Silver		2.5%	
Spongolite		5%	

Talc	Amount B	
Tantalum		The rate is – (a) in the period beginning on 1 January 2003 and ending on 30 June 2003 – (i) 3.3% of the royalty value if sold as concentrate; (ii) 3.3% of the value in concentrate form if processed further before sale; and (b) in the period beginning on 1 July 2003 and ending on 30 June 2004 – (i) 4.1% of the royalty value if sold as concentrate; (ii) 4.1% of the value in concentrate form if processed further before sale; and (c) on or after 1 July 2004 – (i) 5% of the royalty value if sold as concentrate; (ii) 5% of the value in concentrate form if processed further before sale.
Tin		2.5% of the royalty value of tin metal when sold in that form; or, when sold in any other form, 2.5% of the value of the contained tin calculated at the ruling price of tin metal used for the purpose of the sale.
Uranium		The rate is 5% of the royalty value if sold as a uranium oxide concentrate.

(*Continued*)

(Continued)

	Column 1	*Column 2*	*Column 3*
Mineral	*Amount per tonne according to quantity produced or obtained*	*Percentage of the royalty value*	*The rate as specified hereunder*
Vanadium			The rate is – (a) if sold as a concentrate (vanadium oxide), 5% of the vanadium pentoxide price; or (b) if sold in metallic form (ferrovanadium), 2.5% of the ferrovanadium price; or (c) for vanadium not realised on contained vanadium from a product (such as magnetite) where the average grades of vanadium are over 0.275% V_2O_5 in the ore and a vanadium circuit is not installed – 5% of the vanadium pentoxide price.
Zinc			The rate is – (a) if sold as a concentrate, 5% of the royalty value; or (b) if sold in metallic form, 2.5% of the royalty value.
Zircon		5%	
A mineral that is listed in this table when it is in a form that is not specifically listed in this table or any other mineral not specifically listed in this table (excluding gold metal as defined in regulation 86AA and ilmenite feedstock as defined in regulation 86AC)			The rate is – (a) if sold as crushed or screened material, 7.5% of the royalty value; or (b) if sold as a concentrate, 5% of the royalty value; or (c) if sold in metallic form or a form of equivalent processing, 2.5% of the royalty value.

Source: www.dmp.wa.gov.au/Minerals/Royalties-1544.aspx

of the 10 percent royalty rate of the well-head value. If the royalties exceed this 10 percent, the additional amount is attributed to the State (www.dmp.wa.gov.au/Petroleum/Royalties-1578.aspx).

Currently, the following three systems are used for collecting petroleum royalties:

- Well-head system
- Resource Rent Royalty system (RRR)
- Petroleum Resource Rent Tax (PRRT).

The well-head royalty system is based on the gross value of petroleum recovered less all costs incurred between a defined value in the Christmas tree and the point of sale. These costs relate to the processing, storage and transport of the petroleum to the point of sale, and do not include costs of exploration, drilling, recovery and abandonment. Other allowable deductions include depreciation on commissioned post well-head assets and the cost of borrowing on commissioned well-head assets.

Royalty rates of 10% and 12.5% of well-head value generally apply for primary and secondary licenses respectively.

The Resource Rent Royalty (RRR) system was devised for the Barrow Island operation as an incentive to recover the optimal amount of oil. It replaced the well-head royalty that had previously applied (i.e. prior to 1982) and is based on a percentage of net cash flow. Its key aspects are as follows:

- All allowable expenditure, both current and capital, is written off when incurred.
- Exploration costs incurred more than a year prior to the application of the RRR are not allowable deductions.
- Any excess of costs over revenues is carried forward and compounded at a threshold rate.
- Any excess income over the threshold rate is charged to RRR at a rate of 40%.
- It is a tax before income tax and is therefore deductible against income tax.
- The revenue is shared 75% to the Commonwealth and 25% to the State.

The Petroleum Resource Rent Tax (PRRT) is a tax based on profitability applied to all petroleum products (crude oil, natural gas, liquefied petroleum gas and condensate) but not to value-added products including liquefied natural gas. The PRRT is applied at a rate of 40 percent of a project's taxable profit and further details can be found at www.legislation.gov.au/Details/C2012A00018.

Royalty systems in New South Wales (NSW)

When titles for the right to mine and dispose of minerals and petroleum are granted in NSW, Australia, the titleholders accept the responsibility for calculating the amount to be paid and for the completion and delivery of the obligatory returns and payments on time. In NSW, royalties are classified as mineral (non-coal), coal or petroleum royalties.

Mineral (non-coal) royalties

In this sector the following two systems are used:

- Quantum royalty
- Ad valorem royalty.

Quantum royalties are based on a flat rate charge per tonne or unit of mineral produced, and the specific charge depends on the mineral being mined. It is generally used for low value minerals including clay, gypsum and limestone. The following rates apply to the various minerals in NSW:

Mineral	*Rate of Royalty (as at Jan 2018)*
Agate	4% ex-mine value
Agricultural lime	35 cents per tonne
Antimony	4% ex-mine value
Apatite	4% ex-mine value
Arsenic	4% ex-mine value
Asbestos	4% ex-mine value
Barite	70 cents per tonne
Bauxite	35 cents per tonne
Bentonite (& Fuller's Earth)	0 cents per tonne
Beryllium minerals	4% ex-mine value
Bismuth	4% ex-mine value
Borates	70 cents per tonne
Cadmium	4% ex-mine value
Caesium	4% ex-mine value
Calcite	40 cents per tonne
Chalcedony	4% ex-mine value
Chert	35 cents per tonne
Chlorite	70 cents per tonne
Chromite	4% ex-mine value
Clay/shale	35 cents per tonne
Cobalt	4% ex-mine value
Columbium	4% ex-mine value
Copper	4% ex-mine value

Mineral	*Rate of Royalty (as at Jan 2018)*
Corundum	4% ex-mine value
Cryolite	4% ex-mine value
Diamond	4% ex-mine value
Diatomite	70 cents per tonne
Dimension stone	70 cents per tonne
Dolomite	40 cents per tonne
Emerald	4% ex-mine value
Emery	4% ex-mine value
Feldspathic materials	70 cents per tonne
Fluorite	70 cents per tonne
Galena	4% ex-mine value
Garnet	4% ex-mine value
Geothermal substances	4% ex-mine value
Germanium	4% ex-mine value
Gold	4% ex-mine value
Graphite	4% ex-mine value
Gypsum	35 cents per tonne
Halite (inc. solar salt)	40 cents per tonne
Ilmenite	4% ex-mine value
Indium	4% ex-mine value
Iron minerals	4% ex-mine value
Jade	4% ex-mine value
Kaolin	70 cents per tonne
Lead	4% ex-mine value
Leucoxene	4% ex-mine value
Limestone	40 cents per tonne
Lithium	4% ex-mine value
Magnesite	70 cents per tonne
Magnesium salts	40 cents per tonne
Manganese	4% ex-mine value
Marble	70 cents per tonne
Marine aggregate	4% ex-mine value
Mercury	4% ex-mine value
Mica	70 cents per tonne
Mineral pigments	70 cents per tonne
Molybdenite	4% ex-mine value
Monazite	4% ex-mine value
Nephrite	4% ex-mine value
Nickel	4% ex-mine value
Niobium	4% ex-mine value
Oil shale	4% ex-mine value
Olivine	70 cents per tonne

(*Continued*)

(Continued)

Mineral	*Rate of Royalty (as at Jan 2018)*
Opal	4% ex-mine value
Ores of silicon	4% ex-mine value
Peat	70 cents per tonne
Perlite	70 cents per tonne
Petroleum	10% of well-head value
Phosphates	70 cents per tonne
Platinum group minerals	4% ex-mine value
Platinum	4% ex-mine value
Potassium minerals	70 cents per tonne
Potassium salts	40 cents per tonne
Pyrophyllite	70 cents per tonne
Quartz crystal	4% ex-mine value
Quartzite	70 cents per tonne
Rare earth minerals	4% ex-mine value
Reef quartz	70 cents per tonne
Rhodonite	4% ex-mine value
Rubidium	4% ex-mine value
Ruby	4% ex-mine value
Rutile	4% ex-mine value
Sapphire	4% ex-mine value
Scandium and its ores	4% ex-mine value
Selenium	4% ex-mine value
Serpentine	70 cents per tonne
Sillimanite-group minerals	70 cents per tonne
Silver	4% ex-mine value
Sodium salts	40 cents per tonne
Staurolite	70 cents per tonne
Strontium minerals	4% ex-mine value
Structural clay	35 cents per tonne
Sulphur	4% ex-mine value
Talc	70 cents per tonne
Tantalum	4% ex-mine value
Thorium	4% ex-mine value
Tin	4% ex-mine value
Topaz	4% ex-mine value
Tourmaline	4% ex-mine value
Tungsten and its ores	4% ex-mine value
Turquoise	4% ex-mine value
Vanadium	4% ex-mine value

Ad valorem royalties apply to all other minerals as a percentage of the total value of the minerals recovered, and the current rate is 4% of the ex-mine value of production. In calculating the ex-mine value, the processing and treatment costs to produce a saleable product are allowable deductions, whereas the costs associated with exploration, development, mining (production costs) and rehabilitation are not.

Coal royalties

Royalties have to be paid on all coal mined in NSW, and the following types of royalties are currently used:

- Ad valorem royalty
- Coal reject royalty.

The ad valorem royalty for coal is charged as a percentage of the value of production, being total revenue less certain allowable deductions. The coal ad valorem royalty rates are currently:

- 6.2% for deep underground mines (where the depth exceeds coal being extracted at least 400 metres below surface);
- 7.2% for less-deep underground mines; and
- 8.2% for open cut/surface mines.

The coal reject royalty is applicable to coal disposed of for the purpose of producing energy. Coal reject is defined as the by-product of the mining or processing of coal with an energy value of less than 16 gigajoules per dry tonne or contains more than 35% ash by dry weight. The royalty on such coals is no more than 50% of the ad valorem coal royalty.

Petroleum royalties

Currently, a royalty is paid at 10% of the well-head value of the petroleum, except that for titles granted or renewed after 21 August 1992 under the Petroleum (Onshore) Act 1991, the royalty rate for the first five years of commercial production is zero, for the sixth year 6%, and increasing by 1% per annum thereafter up to a maximum of 10% of the well-head value in the 10th year.

The well-head is defined as that point where the petroleum reaches the surface, and the well-head value is the revenue and/or savings from the generation of electricity after deducting costs incurred downstream of the well-head.

Pricing and contract systems

In reviewing this subject matter, Forbes (1980) classified the main pricing systems as follows:

- **Open market pricing**: applicable to the sale of monetary and base metals and, to a lesser extent, aluminium and nickel;
- **Producer pricing**: applicable to those metal markets where there are only a few producers;
- **Published price quotations**: used in the sale of minor minerals and metals where the value of trading is small, e.g. antimony, wolfram (tungsten), bismuth;
- **Contract pricing**: used primarily in the pricing of iron ore, coal, uranium and other non-metallic minerals;
- **Controlled prices**: applicable when there is direct intervention in the markets by governments or government-backed cartels; and
- **Spot market sales**: used for the immediate sale of commodities at existing spot market prices.

Open market pricing

The open market pricing system (www.lme.com) is best displayed by the London Metal Exchange (LME), which was established in 1882 as a clearing house for the world's surplus metals. It organises trade in one of fourteen metals categorised as:

- Non-ferrous metals (copper, tin, lead, zinc, nickel and aluminium)
- Ferrous metals (iron and steel, as well as rebar and scrap)
- Precious metals (gold, silver, platinum and palladium)
- Minor metals (molybdenum and cobalt).

Trading sessions occur each day through LME Select (electronic), the Ring (open outcry) and the 24-hour telephone market, where dealers make bids and offers and deals are immediately concluded on acceptance.

All LME contracts are traded in parcels referred to as lots. Each lot, priced in US$, varies in size from 1 to 65 tonnes depending on the underlying metal and the contract type.

The LME provides an accurate indicator of trends in the world's supply of and demand for various metals. Trading that is not transacted through the LME usually uses prices set by the market.

Other than for immediate purchases and sales, the LME is also used extensively for hedging, options and futures trading, with dealers buying and selling for delivery on any market day between the current day (cash) or three months forward (three months). In the case of silver, the position can be held for up to seven months forward.

The LME indicators of market conditions are the:

- Price levels
- Difference between cash and three-month prices
- LME stocks.

The difference between cash and three-month prices is both indicative of current supply and the market's perception of supply and demand for the short-term future.

If there is plentiful supply of the metal the forward price is normally at a premium over cash, which is a condition known as a contango.

If supplies are scarce the cash price may be higher than that for a three-month sale, and this market condition is known as backwardation.

Large stocks in the LME warehouse are indicative of a stable market, whereas depleted stocks indicate the likelihood of rising prices.

When hedging, buyers and sellers safeguard themselves against fluctuating market prices. For example, if a purchase requires a supply of metal in three months, the requisite amount of three-month metal could be bought on the LME. When the supply date arrives, the buyer will close the hedge by selling his LME contract for cash, so that any loss will be offset by the profit on the paper hedge.

If, on the other hand, a mining company wishes to sell its output to a customer (e.g. a smelter) in three months, it could make a forward LME sale. When the transaction date arrives the company will sell its shipment and close the hedge by buying cash-metal on the LME to make delivery for the forward sale. If the price has fallen in the interim, the loss will be offset by the paper profits generated by the hedge.

There are also bullion markets in London including ATS Bullion (www.atsbullion.com) and an Australian bullion market being AINSLIE Bullion (www.ainsliebullion.com.au). Gold and silver are readily traded through these platforms.

The various exchanges provide a price fixing mechanism and allow for trading and hedging of the commodities.

Another example of open market pricing is provided by the Straits Tin Market, which was established in Kuala Lumpur, as a physical market for Malaysian producers. The price to be paid to producers is determined daily by the Straits Trading Company Ltd. and the Eastern Smelting Company Ltd., which operate their own smelters.

Producer pricing

In some instances, producers of a metal commodity may agree to maintain a standard selling price for some time in order to secure its market against competitive materials. In other instances major producers have used their market dominance to keep metal prices at a low level in order to secure and expand their market share.

Published price quotations

A number of journals publish information on metal sales, and these quoted prices have become, in many instances, the standard prices for minerals and metals transactions. *Metals Week* publishes daily metals prices based on the actual sales of leading sellers, and the quoted prices are the weighted average of all reported sales on that day. These quotations are often used as the basis for the contractual prices of various ores and metals.

Contract prices

Long-term contracts provide the mineral producer with security of demand at a contract price, thereby ensuring the viability of the mine. Such contracts are preferred by the Japanese buyers and Australian suppliers of coal and iron ore.

Clauses allowing price negotiations are included in the contracts and the following list reflects some of the items covered:

- Schedule of sales tonnages
- Options to vary tonnages
- Quality specifications
- Sales price and arrangements for renegotiation
- Price adjustments for quality variations
- Rights of buyers to reject shipments
- Terms and conditions of payments
- Currency of payment and exchange rate provisions
- Force majeure clause
- Arbitration of disputes.

The force majeure clause provides temporary relief from contractual obligations when either the buyer or the seller cannot fulfil its obligations due to unforeseen circumstances, such as acts of God (often associated with adverse weather or significant geological disturbances) and industrial disputes.

It is worth noting that there is a United Nations convention on contracts (www.uncitral.org/uncitral/en/uncitral_texts/sale_goods/1980CISG.html) for the international sale and purchase of goods which could assist in the establishment of contracts and the settlement of disputes. The Centre for International Dispute Settlements (CIDS) provides education as well as a platform for exchanging ideas and research into dispute settlement.

With local contracts, both prices and tonnages tend to be renegotiated annually. Australian coals are usually sold to overseas buyers on the basis of free on board (FOB) prices, which is the value of the coal after deducting all operating and transport costs in getting the coal onto a ship. It will therefore exclude insurance and ship freight costs, as well as loading and unloading costs. Sometimes the price is broken into two components, the ex-mine or free on rail (FOR) price and the remainder, comprising government freight charges and royalties which may be passed on in full to the buyer.

Some coal contracts can also be based on a Free On Board and Trimmed (FOBT) price, which includes the cost of boarding and trimming the coal in the ship's hold.

Controlled prices

Controlled pricing occurs when there is direct intervention in the markets by government or government supported cartels. Perhaps the best example is the International Tin Council, which was formed by the largest producing countries and most of the consuming countries.

The Council aims to avoid surplus production by controlling production and judiciously using buffer stocks to minimise price fluctuations.

A series of international agreements have been promulgated which establish floor and ceiling prices based on LME cash quotations and the financing of a buffer stock. Within the floor and ceiling price range is a neutral zone, and when the tin price is in this zone no action is required by the stock manager. If, however, the price drops below the neutral zone, the stock manager can buy on the LME to support the market. The converse applies if the price increases above the zone.

A more obvious example of controlled pricing is achieved through the Organisation of Petroleum Exporting Countries cartel (OPEC) for oil prices. OPEC manages global oil and petroleum prices by controlling its member-countries' oil production output and sales. OPEC's five founding members, registered in September 1960 in Baghdad, are Iran, Iraq, Kuwait, Saudi Arabia and Venezuela. In addition to these five founding member countries, additional members now include Qatar, Indonesia, Libya, the United Arab Emirates, Algeria, Nigeria, Ecuador, Gabon and Angola.

Spot sales

Minerals and metals producers can, if they wish, sell their commodities on the spot market. A spot contract is a sale of a commodity from a source having the requisite amount of the mineral or metal immediately available for sale.

Gold producers, for example, adopt a sales strategy based partly on forward sales to cover all or a major part of their mining and treatment costs, with the remainder sold on the spot market.

Coal can also be sold on the spot market, with the United States, Indonesia and South Africa being the major players. The spot coal market provides a range of valuable services, including:

- Balancing marginal supplies and demands;
- Providing a mechanism for strategic stock adjustments; and
- Providing benchmark prices for use as a basis for long-term contracts.

References

Boadway, R. and Keen, M. 2014. *Rent Taxes and Royalties in Designing Fiscal Regimes for Non-Renewable Resources*. CESifo Working Paper No 4568. Category 1, Public Finance.

Bradley, P.G. 1986. *Report of the Mineral Revenues Inquiry in Regard to the Study Into Mineral (Including Petroleum) Revenues in Western Australia*, Volumes 1 and 2.

Forbes, V. 1980. *Will It Make a Quid? Some Aspects of the Economics of Mining, Mining in Society Seminar*. University Minerals Industry Advisory Council, Brisbane, August 1980.

Harries, K.J.C. 1996. *Mining Royalties Between Private Parties, Centre for Resource Studies*. Queen's University, Ontario, Canada.

Information Series No. 10: Mining Act 1978 Mineral Royalty and Production Reporting Provisions, Department of Minerals and Energy, WA, March 1995.

4 Accounting for the extractive industry

Accounting for the extractive industries

Extractive industries are defined as those industries involved in finding and exploiting wasting natural resources (minerals or oil and gas), which are those resources that cannot be replaced in their original state by human beings. It follows, therefore, that the term extractive industries embraces all mining, quarrying and oil and gas enterprises.

Due to the specialised nature of the extractive industry, the Accounting Standards Review Board introduced AASB 1022 *Accounting for the Extractive Industries* in October 1989. This standard was then superseded by AASB 6 *Exploration for and Evaluation of Mineral Resources* which became applicable to annual reporting periods beginning on or after 1 January 2005. In contrast to AASB 1022, AASB 6 deals solely with the accounting treatment of exploration and evaluation expenditures. Entities engaged in other phases of extractive operations are required to apply other Australian Accounting Standards. AASB 6 incorporates IFRS 6 *Exploration for and Evaluation of Mineral Resources* issued by the International Accounting Standards Board (IASB). For-profit entities complying with AASB 6 also comply with IFRS 6. Table 4.1 identifies other accounting standards that may apply to other phases of the extractive industry.

Nature of the extractive industries

The extractive industries have the following unique accounting issues:

- Mining and petroleum companies need to explore for and develop mineral resources or oil and gas fields.
- The mineral or oil/natural gas resource or reserve, which is the prime asset of mining and petroleum companies, is a wasting asset.

Table 4.1 Application of other accounting standards to extractive industry stages other than treatment of exploration and evaluation expenditures

Phase of operation/transaction or event	*Relevant accounting standard*
Activities that precede exploration for and evaluation of mineral resources	Framework AASB 116 *Property, Plant and Equipment*, AASB 138 *Intangible Assets*
Development and construction costs	AASB 116 *Property, Plant and Equipment*, AASB 138 *Intangible Assets*
Amortisation of capitalised costs	AASB 116 *Property, Plant and Equipment*
Inventories	AASB 102 *Inventories*
Revenue recognition	AASB 118 *Revenue*
Restoration costs	AASB 137 *Provisions, Contingent Liabilities and Contingent Assets*, AASB 116 *Property, Plant and Equipment*, Interpretation 1 *Changes in Existing Decommissioning, Restoration and Similar Liabilities*

- Most mineral and petroleum exploration activities are unsuccessful and as such activities are high-risk. These risks should be reflected in the information disclosed in the financial statements.
- The pre-production activities are capital intensive and there can be a long lead time between discovery and the establishment of full production. Given the long lead-time to establishing and extracting an economic resource, rapid changes in metal or petroleum prices may make a previously economic resource unviable to extract.
- There is a lack of certainty regarding the income to be generated from the sale of mineral commodities due to its dependence on supply and demand in world markets.

Typically pre-production stages include:

1 **Exploration**: The search for an economic mineral deposit or an oil or natural gas field.
2 **Evaluation**: Assessing the technical and technological feasibility and commercial viability of the proposed exploitation of the mineral deposit or petroleum field.
3 **Development**: Establishing access to the mineral deposit or petroleum reservoir.
4 **Construction**: Establishing and commissioning facilities for the extraction, treatment and transportation of the product.

Whereas these stages appear to be well compartmentalised, there are some problems in defining the boundaries between them, such as:

- Exploration and evaluation can proceed concurrently so that the allocation of funds is not clear cut.
- The commencement of production, which follows the construction stage, can be defined in various ways, such as:
    - Initial commencement of production;
    - Achievement of commercial level of production; and
    - Achievement of designed level of production.

Accounting treatment of exploration and evaluation expenditures

AASB 6 is limited to the accounting treatment of exploration for and evaluation of mineral resources. The Standard does not address other aspects of accounting by entities engaged in extractive related activities other than exploration for and evaluation of mineral resources. For instance, AASB 6 does not apply to activities prior to the exploration for and evaluation of mineral resources such as costs incurred relating to preliminary reviews of the mineral or petroleum potential of a given geological area which are typically deemed to be prospecting in nature. Additionally, AASB 6 does not apply to activities (e.g. development, mining, and rehabilitation) after an entity has determined that the mineral resource is economically viable to extract.

Area of interest method

It should also be emphasised that many pre-production costs relate to activities whose outcome will not be known for some time, e.g. costs associated with the exploration on a mineral lease or within a petroleum field.

AASB 6 requires entities to apply 'area of interest' accounting to their exploration and evaluation expenditures. The area of interest refers to an individual geological area or a lease within which it is considered favourable for the occurrence of mineral resources or petroleum resources. An area of interest may consist of the area enclosing an individual mineral resource or mine/petroleum field or several resources concentrated within a particular area within which an entity has conducted exploration and evaluation.

According to paragraph AUS7.1, AASB 6, for each area of interest, expenditures incurred in the exploration for and evaluation of mineral resources shall be expensed as incurred, or partially or fully capitalised with those costs being recognised as exploration and evaluation assets if the following conditions (as outlined in paragraph AUS7.2) are met for each area of interest:

a) The rights to tenure of the area of interest are current; and
b) At least one of the following conditions are met:

    1 The exploration and evaluation expenditures are expected to be recouped through successful development and exploitation of the area of interest, or alternatively, by its sale; and
    2 Exploration and evaluation activities in the area of interest have not at the reporting date reached a stage which permits a reasonable assessment of the existence or otherwise of mineral reserves, and active and significant operations in, or in relation to, the area of interest are continuing.

The ultimate recoupment of costs capitalised for exploration and evaluation is dependent on successful development and commercial exploitation of commodities, or the sale of specific areas of interest that contain the resources or reserves. In summary, exploration and evaluation expenditure is recognised in the income statement as incurred, unless the expenditure is expected to be recouped through successful development and exploitation of reserves in an area of interest, or alternatively by its sale, in which case it is recognised as an asset on an area of interest basis. Once commercial viability of the extraction of reserves in an area of interest is determined, exploration and evaluation assets attributable to that area of interest are then reclassified to property and development assets within property, plant and equipment category.

Internationally, other approaches to account for pre-production costs may be employed by entities. These include the full cost method and the successful efforts method. Under the full cost method, all costs associated with exploration, acquisition and development of resources are capitalised for matching against revenue generated from the sale of minerals or petroleum products. This method tends to be used by smaller petroleum companies but less so by mining companies. Under the successful efforts method, pre-production costs that result in the exploitation of resources are carried forward. If, however, exploration does not result in discovery of economic resources, then those costs are expensed. Australian resource entities apply the area of interest method and not these other methods when accounting for pre-production costs.

Recognition and measurement of exploration and evaluation assets

According to paragraph 8, AASB 6, exploration and evaluation assets shall be measured at cost at recognition. Based on paragraph 9, the following are examples of expenditures which might be included in the initial measurement of exploration and evaluation assets:

a) Geological, geochemical and geophysical surveys and sampling;
b) Drilling and trenching;
c) Technical and commercial feasibility studies involved in evaluating a mineral resource.

According to paragraph AUS9.1, AASB 6, both direct and indirect costs are allocated to an area of interest where they are associated with the exploration for and evaluation of mineral resources. The costs of acquiring leases or the right to explore are included in the cost of the exploration and evaluation asset given that they are a necessary part of the exploration for and evaluation of mineral resources. Indirect costs such as depreciation relating to equipment used on an exploration lease are included in the cost of an exploration and evaluation asset. General and administrative costs are included in the cost of exploration and evaluation assets where they are directly related to those assets within an area of interest otherwise they are to be expensed as incurred.

Historical cost versus fair value

Most items of property, plant and equipment employed as exploration and evaluation assets will be measured based on their historical cost in accordance with AASB 116. After initial recognition, an entity shall apply either the cost model or revaluation model to exploration and evaluation assets using the model in AASB 116, or AASB 138 *Intangible Assets*. Paragraph 16 of AASB 6 states that an entity shall classify exploration and evaluation assets as tangible (e.g. exploration equipment) or intangible (e.g. rights to explore) according to the nature of the assets. Historical cost information as well as fair value information may be not always generate useful information for users of financial reports because they may not be useful in future cash flows expected to be generated from mining or oil and gas properties. Historical cost information may faithfully represent the initial cost of exploration and evaluation assets but may not be relevant in valuation of mining properties or mineral or oil and gas resources and reserves.

Accounting policies and accounting estimates

Taylor et al. (2012) explains that resource firms have diverse accounting policies relating to the recognition of pre-production costs, reserves and resources (e.g. historical cost or fair value), the timing of revenue recognition and their associated disclosures. The resulting diversity of accounting policies can give rise to uncertainty regarding treatment of key costs. Paragraph 6 of AASB 6 requires that entities recognising exploration and evaluation assets apply paragraph 10 of AASB 108 *Accounting Policies, Changes in Accounting Estimates and Errors* and paragraphs AUS7.1 and AUS7.2 of AASB 6. In other words, management should exercise its judgement in developing and applying an accounting policy that is relevant and reliable to the decision-making of users. In regards to the extractive industry, the accounting policy is to employ an 'area of interest' approach which provides the basis of determining whether expenditures incurred are to be expensed or capitalised. AASB 6 (in paragraph 7) exempts an entity from consideration of authoritative guidance as specified in paragraphs 11 and 12 of AASB 108 *Accounting Policies, Changes in Accounting Estimates and Errors* in the development of an accounting policy, and instead, entities can use standard industry accounting practices in the recognition and measurement of exploration and evaluation expenditure.

The preparation of financial statements requires management to make judgements and to make assumptions when deriving accounting estimates. Application of different assumptions may have a significant impact on the net asset position and profitability of resource entities. For example, areas which involve a high degree of complexity and judgement or areas where assumptions may have a significant impact on financial statements include, for example, reserve estimates, exploration and evaluation expenditure, development expenditure, property, plant and equipment-recoverable amount, and rehabilitation estimates.

Development expenditure

In accordance with paragraph 10 AASB 6, expenditures related to the development of mineral resources shall not be recognised as exploration and evaluation assets. Such expenditure instead are to be capitalised as development or production property and accounted for under AASB 116 *Property, Plant and Equipment*. Exploration and evaluation assets are not depreciated given that they are not available for use and are instead subject to impairment testing in accordance with AASB 136.

Accounting for mineral resources and reserves

The Joint Ore Reserve Committee Code (JORC Code) sets out guidelines for public reporting in Australia of exploration results, mineral resources and ore reserves (The JORC, 2012). Exploration results include data and information generated by mineral exploration programmes that might be of use to investors. A mineral resource is a concentration or occurrence of solid material of economic interest. Public reporting (e.g. annual reports) of mineral resources must satisfy the requirement that there are reasonable prospects for eventual economic extraction (i.e. more likely than not), regardless of the classification of the mineral resource. An ore reserve is the economically mineable part of a measured and/or indicated mineral resource. Resource and reserve assets account for a significant part of the equity value and profits of resources firms (Taylor et al., 2012). In fact, the disclosure of resource and reserve valuations and mine life or petroleum reservoir parameters are important in forecasting earnings, cash flows and growth potential for resource firms and may therefore be a significant determinant of their share prices. However, there is no Australian accounting standard that mandates disclosure of reserve parameters.

The Corporations Law in Australia does not require disclosure of the quantity or value of mineral resources or reserves. The Australian Stock Exchange, on the other hand, states that a company report on the progress of any geological survey must be in accord with the Australian Institute of Mining and Metallurgy (AIMM) Code (i.e. the JORC Code) for Reporting of Identified Mineral Resources and Ore Reserves. However, this only applies to the information submitted to the exchange and does not refer to the information presented in the annual financial report. It follows, therefore, that disclosure of the quantity and value of mineral resources and reserves in an entity's annual financial report is entirely voluntary. Ideally, the inputs used to calculate mineral reserves should be disclosed including tonnages, grades, discount rates and metal prices. The reason for this is that mineral reserves are the major assets of extractive industry entities. Additionally, mineral reserve values provide the basis for predicting future cash flows, earnings and share values. Financial statements based on reserve values would, therefore, be much more informative than those based on historical cost accounting.

Impairment testing

AASB 6 (paragraph 18) requires entities recognising exploration and evaluation assets to perform an impairment test in accordance with AASB 136 *Impairment of Assets* on those assets when facts and circumstances suggest that the carrying amount of an asset may exceed its recoverable amount. This standard also permits impairment of exploration and evaluation assets to be assessed at a cash-generating unit or group of cash-generating units level. The level is not to be larger than the lesser of an area of interest or a segment (based on AASB 114 *Segment Reporting*).

Paragraph 17 AASB 6 requires entities to conduct an impairment test, prior to reclassification of exploration and evaluation assets to property and development assets when technical feasibility and commercial viability of extracting a mineral resource are demonstrable.

Depreciation and amortisation

Mineral resources or petroleum reservoirs are wasting assets in that their useful lives are determined by the quantum of mineral or petroleum reserves. Typically, equipment or plant used to extract and process mineral or petroleum reserves are depreciated using the unit-of-production (UoP) depreciation method.

Under the UoP depreciation approach, depreciation charged for a period = number of units produced in the current period x UoP rate.

The UoP rate = cost of equipment/plant – residual value/estimated number of units to be produced by the equipment/plant over its estimated useful life.

Restoration expenditure

In accordance with paragraph 11 AASB 6, exploration for mineral or petroleum resources often involves some degree of environmental disturbance. In accordance with AASB 137 *Provisions, Contingent Liabilities and Contingent Assets*, an entity recognises any obligations for removal and restoration that are incurred as a consequence of having undertaken exploration for and evaluation of mineral resources. Provisions may be made for such things as plant closure, mobilisation of oil rigs, decontamination, regeneration of vegetation and ongoing monitoring. Restoration work often is undertaken progressively within an area of interest and hence provisions are carried forward. Although provisions are re-assessed every period, there is often considerable uncertainty with estimating the amount of the provision as this will depend on factors such as changes in environmental legislation, changes in technology, changes in expected future costs and changes in future interest rates.

Disclosures

According to paragraph 23 of AASB 6, an entity shall disclose information that identifies and explains the amounts recognised in its financial report arising from the exploration for and evaluation of mineral resources. In accordance with paragraph 24 AASB 6, an entity shall disclose:

a) Its accounting policies for exploration and evaluation expenditures including the recognition of exploration and evaluation assets; and

b) The amount of assets, liabilities, income and expense and operating and investing cash flows arising from the exploration for and evaluation of mineral resources.

Additionally, an entity is to provide an explanation that recoverability of the carrying amount of exploration and evaluation assets in respect to an area of interest is dependent on successful development and commercial exploitation, or alternatively, sale of that area of interest.

An entity shall treat exploration and evaluation assets as a separate class of assets and make disclosures required by either AASB 116 or AASB 138 consistent with how the assets are classified.

Financial risk management

Extractive industry companies are exposed to a range of financial risks, including market risk (commodity, interest rate and foreign currency risks), credit risk and liquidity risk. In particular, mining companies are exposed to the commodity price risk, given that mineral sales are predominantly subject to prevailing market prices. Mining companies have limited ability to directly influence market prices of minerals and typically manage commodity price risk through focus on improving cash margins, refinancing and through use of contracts to hedge or lock-in the prices at which they can sell minerals in the future.

References

Australian Accounting Standards Board (AASB) Standard 6 *Exploration for and Evaluation of Mineral Resources*. Available at: https://www.aasb.gov.au/Pronouncements/Current-standards.aspx

Australian Accounting Standards Board (AASB) Standard 1022 *Accounting for the Extractive Industries*. Available at: https://www.aasb.gov.au/Pronouncements/Current-standards.aspx

Taylor, G., Richardson, G., Tower, G. and Hancock, P. 2012. Determinants of Reserve Disclosure in the Extractive Industries: Evidence from Australian Firms. *Accounting and Finance*, 52 (1), 373–402.

The JORC Code 2012 Edition. Australasian Code for Reporting of Exploration Results, Mineral Resources and Ore Reserves. Available at: http://www.jorc.org/docs/jorc_code2012.pdf

Printed in the United States
by Baker & Taylor Publisher Services

Printed in the United States
by Baker & Taylor Publisher Services